Selina Gaye

The World's Lumber Room

A Gossip About Some of it's Contents

Selina Gaye

The World's Lumber Room
A Gossip About Some of it's Contents

ISBN/EAN: 9783744717236

Printed in Europe, USA, Canada, Australia, Japan

Cover: Foto ©Thomas Meinert / pixelio.de

More available books at **www.hansebooks.com**

THE WORLD'S LUMBER ROOM.

SOME OF NATURE'S SCAVENGERS.

(See Chap. XIV.

THE
WORLD'S LUMBER ROOM

A Gossip about some of its Contents

BY

SELINA GAYE.

"I regarded myself as one of those vile things that Nature designed
should be thrown by into her *lumber room*, there to perish in obscurity."
Vicar of Wakefield.

"Old decays but foster new creations ;
Bones and ashes feed the golden corn."
Charles Kingsley.

With Fifty-seven Illustrations.

CASSELL & COMPANY, Limited:

LONDON, PARIS, NEW YORK & MELBOURNE.

1885.

PREFACE.

THE object of this volume is to give, in popular form, an account of some of the many ways in which refuse is made and disposed of, first and chiefly by Nature, and secondly by Man.

The World's Lumber Room, comprising the three great departments of Earth, Air, and Water, is in fact co-extensive with the World itself, and, so far from being the sort of place which the worthy Vicar's son seems to have pictured to himself, is rather a workshop or laboratory, where nothing is left to "perish," in his sense of the word, but the old becomes new, and the vile and refuse, instead of being "thrown by" in their vileness, are taken in hand and turned to good account.

Considering the popular character of the volume, it was thought undesirable to cumber the pages with references, but the following are the works which have been chiefly consulted—"Elements of Chemical and Physical Geology," by Gustav Bischof; Darwin's "Journal of Researches," "Coral Reefs," and "Vegetable Mould and Earthworms;" J. D. Dana's "Coral Reefs and Islands;" Sir C. W. Thomson's "Voyage of the *Challenger*;" H. N. Moseley's

"Notes of a Naturalist on the *Challenger;*" Maury's "Physical Geography of the Sea;" Prof. Nordenskjöld's "Voyages;" Huxley's "Physiography of the Thames Valley;" Belt's "Naturalist in Nicaragua;" Bates's "Naturalist on the Amazons;" A. R. Wallace's "Australasia;" Prof. Tyndall's "Fragments of Science;" S. W. Johnson's "How Plants Grow," and "How Plants Feed;" Miss E. F. Staveley's "British Insects;" J. O. Westwood's "Modern Classification of Insects;" Wilson's "American Ornithology;" Lord Dunraven's "Great Divide;" Brachvogel's "Mammoth Springs;" P. L. Simmond's "Waste Products;" and various articles in "Science for All," "Journal of Science," "American Journal of Science," "Quarterly Journal of the Geological Society," and "Gartenlaube," &c., &c.

CONTENTS.

LIST OF ILLUSTRATIONS.

THE WORLD'S LUMBER ROOM.

CHAPTER I.

DUST HO!

THE dustman's cart! Probably ninety-nine persons out of every hundred, if questioned as to its contents, would declare off-hand that it held nothing but *rubbish*—rags, bones, bits of glass and crockery, oyster-shells, paper, cabbage-stalks, ashes, broken things, and what was of no use to any one.

But the hundredth would look at it with different eyes, and give a different answer.

"It is rubbish to you," he might say, "because it is 'matter out of place'; but only follow it to the dust contractor's yard, and see how many people get their living out of this rubbish. A good-sized heap of dust has been known to be worth £4,000 or even £5,000."

And if he were a poet or a scientist, he might go on to assure you that the unsavoury-looking cart was "brimful of possibilities, more wonderful than are dreamt of in fairy land."

B

He might look at "bones and ashes" and tell you that he saw "golden corn;" at old rags, and say, "from cast-off clothes come bitter beer and early cauliflowers."

In fact, he might go through the whole list and prove to you that, so far from it *all* being rubbish, *none* of it is rubbish, when it gets into its right place, and that every single item has a more or less wonderful future before it; for, where man's ingenuity fails, Nature comes to the rescue, and she is intolerant of all waste.

Then again, if archæologically inclined, he might look at the dust and think of the ancient dust-heaps discovered or disinterred in various parts of the world, which tell in such plain language of the habits and manner of life of races long since passed away from the earth, and he might speculate upon the possibility of one of these modern dust-heaps being buried and preserved to become an interesting historical record, telling future generations what the ancient Englishman lived upon, what fruits and vegetables he cultivated, and what he imported, what fabrics he wore, what perfection he had attained in the arts of pottery and of glass- and paper-making, what an extensive trade he must have had with all parts of the world, and many other things besides too numerous to mention.

To the dustman everything which finds its way into his cart, whether it be ashes, old shoes, or an occasional silver spoon—all comes under the head of " dust." But, leaving these for the present, we will limit our inquiries to the fine powdery matter, commonly called *dust*, which is by no means the insignificant, unimportant trifle it at first sight appears. For dust has caused lawsuits, dust has caused shipwrecks, and dust has caused, and still causes, death.

More than this, without dust we should probably have no fogs, no clouds, and no rain, there would be no light, no colour in the ocean ; and finally, without dust, the sun's rays would be invisible to us.

What then is this dust, and where does it come from ?

Inside our houses it consists chiefly of the minute particles rubbed off carpets, furniture, and the like, by the ordinary wear and tear of daily life. Carpets and clothes are constantly being worn threadbare by perpetual rubbing ; and every time we cut the leaves of a book, wind wool, use a sewing-machine, scratch the legs of table or chair, every time, in fact, that we walk across the room, we help to make the dust which the housemaid sweeps up in the morning.

But it is not only the things around us which are being thus continually rubbed and worn. Something of the same sort happens to ourselves. We are perpetually changing our skins, not indeed as the lizard does, but piecemeal ; and, as the little loose particles are rubbed off, they, too, help to make the dust which collects in our houses, and needs to be constantly removed, if our bodies are to be kept in a healthy state.

For the greater part of this dust is organic ; it has at one time or other formed part ot some organised living thing, animal or vegetable, and, as such, it will burn quickly if we set fire to it, or slowly if we spread it upon the fields, and leave it, as we say, "to decay." * It

* See Chap. xii.

cannot therefore be wholesome to have it about us, as, though the exact nature of the germs which spread scarlet fever and other diseases may not be determined, one thing is certain, namely, that they flourish most in the neighbourhood of dirt and decaying matter of all kinds.

The lighter part of this dust floats in the air, as we see whenever a sunbeam shines into the room and reveals it to us; the heavier settles on the floor and furniture, where, in towns at least, it is so blackened by dust of another kind—the smoke and soot arising from the imperfect combustion of fuel—that it is not easy to tell what its original colours were.

Still, the examination of any bit of "flue" which has collected behind some piece of furniture too heavy to be constantly moved, will show us that it consists in the main of fine hairs and particles of wool, worsted, cotton, &c., which have been worn off by the housemaid's broom and our own movements, and felted together by pressure, some being large enough to show their origin plainly, while others are mere dust, scarcely distinguishable as hairs at all by the naked eye.

But besides the dust made inside the house, some of that made outside also finds its way in, and in dry, windy weather covers the furniture with a very unpleasant gritty film; for it is so exceedingly fine that, like very fine snow, it will make its way through the smallest cracks and crevices, and is not to be kept out by closed windows.

In towns, much of this dust also is made by rubbing and friction. The wear and tear occasioned by the incessant passing of feet and the wheels of heavily-loaded vehicles are evident enough ; for the streets are constantly needing to be repaired.

On the large flagstones in the main road between the West India Docks and Whitechapel, for instance, where the traffic is, of course, exceptionally heavy and unceasing, three inches of hard granite were worn away in the course of forty years; and the roads of Funchal, in the island of Madeira, which run up steep slopes and are paved with basalt, have become polished and slippery from constant use.*

Every one has noticed, too, how very dusty railway-carriages are apt to be; and though some of the dust in them consists of particles rubbed from the cushions and curtains, and some of sand from the road, yet a large proportion will adhere to a magnet and prove to be minute fragments of iron and steel worn from the wheels and rails.

But the world's dust is not made solely by man. Nature, too, is for ever making it in a variety of ways ; and as dust of all kinds finds its way into our houses, it is not too much to say that, before we can account satisfactorily for all the contents of the dustman's cart, we must know something of Nature's labourers and their methods of working. The oyster-shells are, of course, indebted to her dust-makers and dust-carriers for the materials of which they are made, and the same frost and heat which have broken countless

---

* Mr. Bates, in his " Naturalist on the Amazons," mentions that those mother turtles which had laid eggs the previous year were easily known by the fact that the horny skin of their breast-plates was worn by their having crawled over the sand.

glasses are likewise busy with the rocks ; but, besides these very obvious items, the fine gunpowder-like dust, which is sifted out from the other refuse, must contain specimens of Nature's dust in most, if not all, of its varieties.

Night and day, summer and winter, her great army is ever at work, cutting, carving, grinding, loosening, polishing,

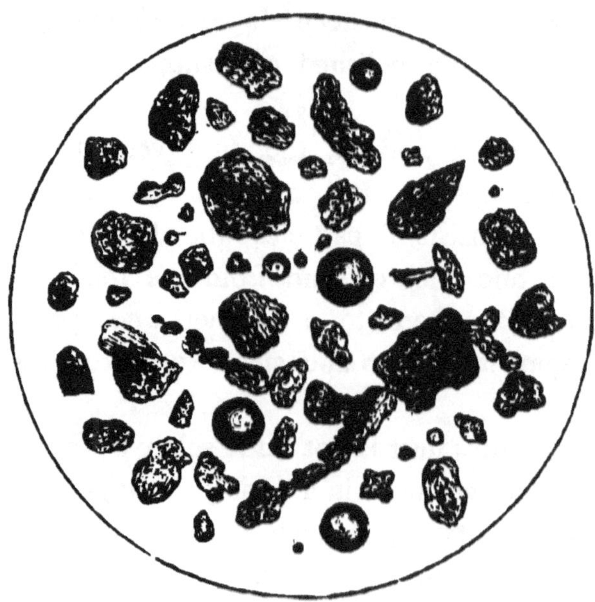

Fig. 1.—COSMIC DUST.

hammering the rocks, and making an impression even on the very hardest of them.   One result of all this wear and tear is the soil which almost everywhere covers the earth's crust, sometimes to the depth only of an inch or two, sometimes even less, while sometimes again it attains a thickness of several feet.   At the utmost, however, its depth may be measured by feet, and at its greatest thickness it is to the earth only as the thin film of dust which settles on our furniture.

But then, again, some of the mineral-dust which finds its way into our streets, and thence into our houses, has been made altogether outside the world, and has floated

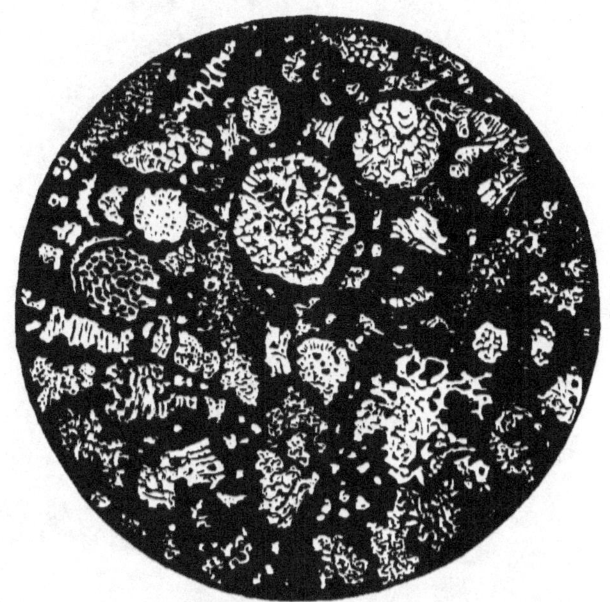

Fig. 2.—MICROSCOPIC SECTION OF METEORIC STONE.

down through the air from regions of which we know nothing more than the little we can learn from this dust and by means of the telescope.

Cosmic or meteoric dust (Fig. 1) is of the same nature as those larger fragments of matter which, when they take fire on coming into contact with the earth's atmosphere, and blaze for a few instants in the sky, we call "shooting" or " falling" stars. The number of these meteorites (Fig. 2) is simply inconceivable, since they form, we are told, a ring one million miles long and one million deep, and a celebrated American astronomer calculates that on an average there fall to the earth every day 7,500,000 of a size to be seen by the naked eye, and 400,000,000 which might be seen through

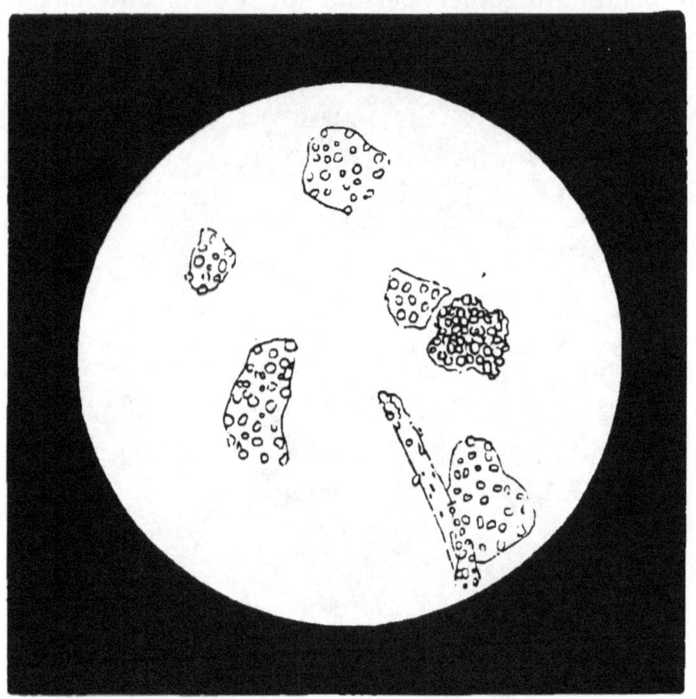

Fig. 3.—ORGANIC MATTER FOUND AFTER THE EVAPORATION OF SNOW
(*Magnified 500 times.*)

a telescope of moderate power, and that, with the fine dust which accompanies them, the daily weight falling upon the earth does not come short of a ton.*

But the observations of Professor Nordenskjöld lead him to go much farther even than this. Where the dust falls upon roads or fields there is of course some difficulty

* Single meteorites sometimes weigh several tons.

in distinguishing it from dust of earthly origin ; but where it falls upon fields of eternal snow and ice (Figs. 3 and 4), far removed from any rock or soil which could produce ordinary dust, it is more easily collected ; and, from his

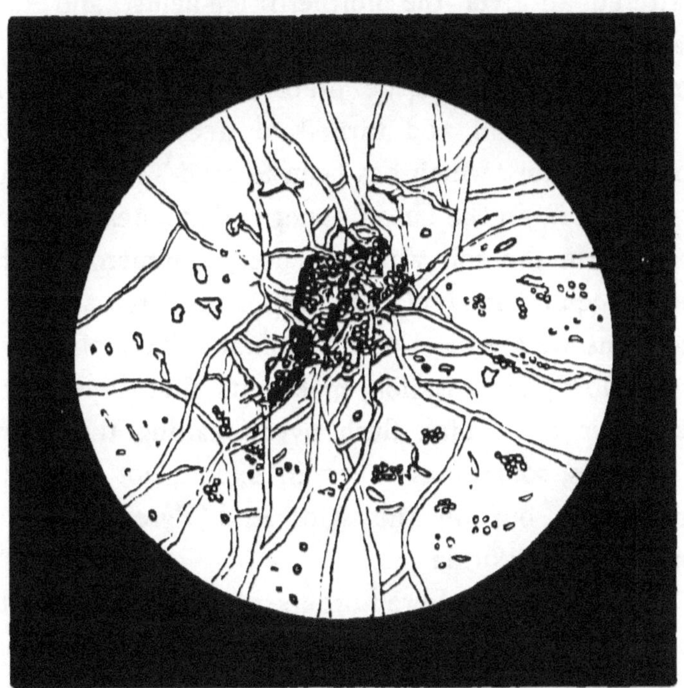

Fig. 4.—VEGETATION IN A DROP OF EVAPORATED SNOW.
(*Magnified 500 times.*)

experience in the Arctic regions, the Professor is of opinion that more than 500,000 tons fall uniformly and steadily over the whole globe in each year.

Observations, since repeated by the Russian scientists in Central Siberia, have yielded similar results, for the dust is found to be composed in large measure of metallic par-ticles, such as are characteristic of meteorites, though others of volcanic origin may, no doubt, also be mingled with them, since we know that fine volcanic dust may be transported almost any distance by currents of air.

The ice-dust or "kryokonit," as it is proposed to call it, is scattered all over the northern ice-fields, and though well aware that it cannot be lost or wasted in the long run, we may hardly, perhaps, be prepared to find that it is at once taken in hand and turned to account. Kryokonit, the meteoric dust which has been formed—who can say by what process?—in the far-off upper regions, descends to the earth to form the soil which nourishes numerous hitherto unknown ice and snow plants, besides the "red snow," with whose name we are most of us familiar. This plant, which is of very lowly organisation, makes its first appearance in the summer as a pink flush overspreading the snow in large patches, sometimes miles in extent, not only in the Polar regions, but on the Alps and Pyrenees and the mountains of California.

The spores of the lower orders of plants are very tenacious of life, and are capable of bearing such extremes of temperature that no climate will destroy them. You may even expose them to a heat of 100° C. (212° Fahr.), or to the lowest degree of cold obtainable ($-$ 100° C.), and still, though lying dormant, they will retain their vitality, and as soon as they have the opportunity will grow and multiply.

Though dwelling among snow and ice, however, they need the summer sun to waken them into active life, and they need more than snow and ice to live upon. Like the green slime of our ponds, they belong to the order of plants called *Algæ*, and the red snow, though so minute as to look like nothing but an assemblage of tiny globules, even when seen under the microscope, is still found, like the plants of higher orders, to contain many minerals.* The outer skin of the globule especially yields flint, nor are lime, iron, and the other minerals essential to plant-life, wanting.

Another alga of a brownish-red colour, though closely related to the red snow, is never found except on the ice, where it grows in the blackish mud of the kryokonit either on the surface or at the bottom of the deep holes which, in the summer, pierce the ice in all directions to the great inconvenience of explorers.

These holes are indeed made by the alga itself, which absorbs more heat than the surrounding ice, thanks to its darker colouring, and thus it melts and sinks deeper and deeper, until it is beyond the reach of the sunbeams. Professor Nordenskjöld even imagines that this microscopic plant may have had the chief share in melting those vast fields of ice which in a former age covered great part of Europe and America.

But the snow- and ice-plants serve other purposes besides this. By feeding on the cosmic dust they convert it into food capable of supporting animal life, and many minute creatures, even in the Polar regions, are nourished by the various, red, green, and brown algæ, while the little black "glacier-flea" lives almost entirely on the red snow and its remains.

In addition to the cosmic dust, the air is at times charged with volcanic dust, the finer particles of which,

* See Chap. vii.

having been carried up to great heights, remain suspended for weeks, and even months. After the great eruptions of Skaptar, in 1783, Iceland was obscured for months by fine dust, which was carried over England and the north of Europe, producing fogs, and lurid sunrises and sunsets.

During the great eruption in Java, which reached its climax at the end of August, 1883, volcanic matter was ejected in enormous quantities, and to a height which it is impossible to determine. Millions of tons of matter were hurled into the air, enveloping the whole district for many miles in utter darkness; and although the heavier particles, of course, soon fell to the earth, quantities of fine dust were carried into the upper regions by the tremendous upward current, which always exists in the neighbourhood of the equator. To what height the air brought near to the equator by the trade-winds is ordinarily carried we do not know, neither do we know certainly what becomes of it, but it frequently travels at the rate of one hundred and fifty miles an hour, and presuming it to have done so in this instance, carrying the volcanic dust with it, it would, says Mr. Norman Lockyer, have reached various parts of the world exactly at the time when attention was first drawn to the wonderful magnificence and duration of the sunset glow, and to other unusual appearances in the sky during the autumn and winter of 1883.

These remarkable sunsets and sunrises began in the Mauritius on the 28th of August, and were at once believed to be caused by sunlight passing through fine dust; and from that time for several months they were constantly observed, now in one place, now in another, until they had been seen almost all over the world, having made their first appearance in England on the 9th of November. Here they were noticed, more or less, for many weeks; and from observations made at Berlin during the last three days of November, it was calculated that the reflecting matter must even then have been suspended at a height of forty miles above the earth.

But the dust which constitutes the chief part of the "red fog" of the Atlantic, the "sea-dust" of the northern seas, and the "sirocco-dust" of South Europe, is neither cosmic nor volcanic, though it, too, has travelled great distances.

Vessels, hundreds of miles from land, have been at times enveloped for days in fog consisting of a brick-red or cinnamon-coloured dust, which covered the sails and rigging with a thick coat, and rendered the air so hazy that no vessel which was more than a quarter of a mile off could be distinguished even at mid-day. Off St. Jago, during certain months of the year, a very fine dust is almost constantly falling, which, says Mr. Darwin, roughens and slightly injures astronomical instruments, hurts the eyes, dirties everything on board, and at times falls so thickly that vessels have been known to run ashore owing to the obscurity, and are recommended to avoid the passage between Cape Verd and the Archipelago. Considerable quantities continued to fall upon the *Beagle* when she was between three and four hundred miles from shore, some of the variously-coloured transparent particles being a thousandth part of an inch square, few larger, and the greater part consisting of fine powder. On the succeeding days, as the vessel proceeded on her way, the dust became so fine that it could be collected only on a damp sponge ; but it has been known to fall on vessels one thousand and even sixteen hundred miles from any land.

From the direction of the wind, and the fact that the dust falls during those months when the harmattan raises clouds of it high in the air, and blows from the north-west shores of Africa, it was at first naturally concluded that the dust was all African too. The mineral part of it, no doubt, may be so ; but on examination it was found that the greater part of the dust was organic, and consisted mainly of those minute, flint-cased forms of plant-life known as diatoms,

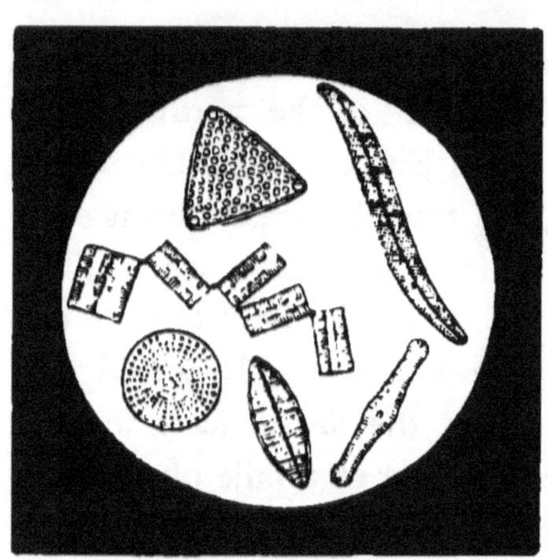

Fig. 5.—DIATOMS.

which exist in almost all water—salt, fresh, or brackish (Fig. 5). Further examination also revealed the singular fact that, though the dust came directly from Africa, of all the many different organic forms none were peculiar to Africa, and all but two belonged to fresh-water families ; and it has since been proved that all the organic portion of the dust, whether it fall at Cape Verd, Malta, Genoa, Lyons, or in the Tyrol, has come from the south side of the equator, and has been transported from the banks of the Orinoco and Amazons.

When, however, we find that particles of mineral matter

one-thousandth part of an inch square can be carried three or four hundred miles, and that a narrow strip of vegetable substance something more than half an inch long, and the twelfth part of an inch wide, clearly belonging to some

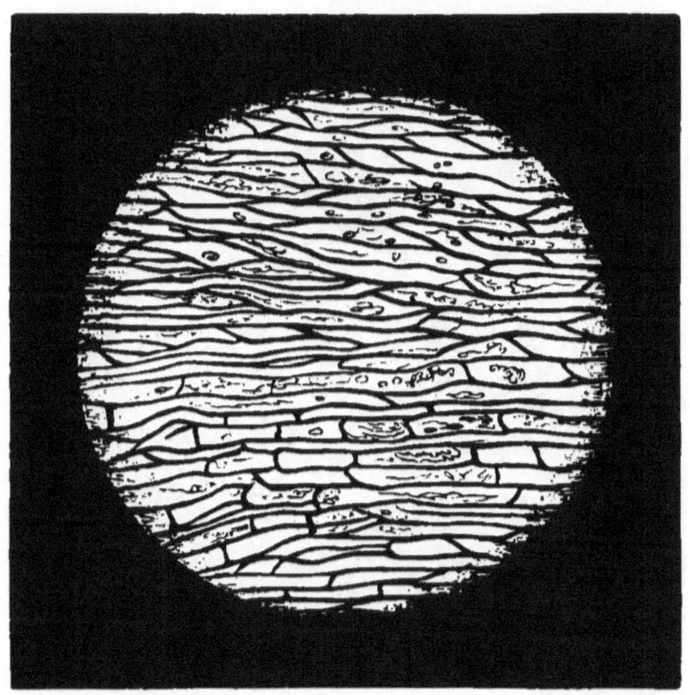

Fig. 6.—VEGETABLE DÉBRIS IN SNOW. (*Magnified 500 times.*)

tropical tree, has been carried more than 1,200 miles from any coast where it could have grown, we wonder less at the long journeys taken by these minute one-celled plants, 41,000,000,000 of which occupy only one cubic inch of space, and weigh but 220 grains.

But we have still to speak of that which may more strictly be called the floating matter of the air, that, namely, which is always present everywhere, and is revealed to us in part, but only in part, whenever a sunbeam shines into the room.

"For the sun," says Daniel Culverwell, "discovers

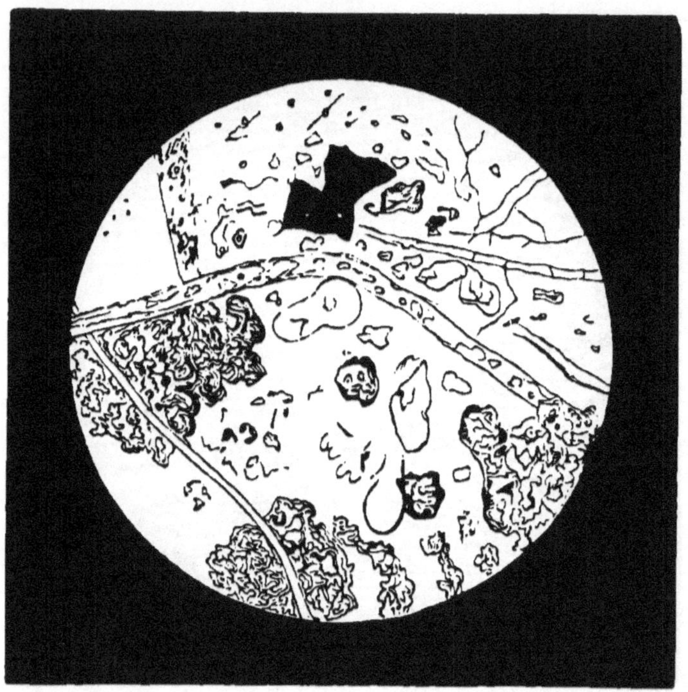

Fig. 7.—VEGETABLE DÉBRIS FOUND IN SNOW. (*Magnified 500 times.*)

atomes, though they be invisible by candle-light, and makes them dance naked in his beams."

And the electric beam does even more, for it shows us that instruments which have been washed, and even polished, and look spotlessly clean in ordinary light, are simply *dirty* when seen by its searching light, if they have been left but an instant exposed to the air.

The greater part of the floating matter of the air is organic, for the heavier mineral dust is winnowed from it, and falls first to the ground. Even the organic matter, light as it is, would fall to the ground also, if the air were perfectly still, which, of course, it never is in Nature.

This floating dust consists of ground-up straw and rags, smoke, seeds, pollen, spores, germs, &c., which invade both air and water to such an extent, that neither, however pure, is entirely free from them (Figs. 6 and 7).

The pollen or fine dust contained in the anthers of flowers and catkins is often wafted into the air in thick clouds in the neighbourhood of pine-forests, where it is known by the popular name of "showers of sulphur."

This is the dust with which the bees powder their coats as they make their way in and out of the blossoms ; and so fine and light is it that much is scattered in this way, and floats away in the air. Then there are the spores which, in flowerless plants—such as ferns, lichens, and fungi,—take the place and answer the purpose of seeds. At the back of the common polypody fern may be seen a number of little round yellow dots, which are often called seeds, though they are not even spores, but spore-cases. Each dot consists of fifty spore-cases, and each case contains thousands of microscopic spores, which, when perfectly ripe and dust-like, are set free to be carried hither and thither by the air.

Then there are the fungus-spores, the largest of which are invisible without the aid of a microscope. These, since they make up in numbers for what they lack in size, are constantly suspended in the air in large quantities, ready to take possession of any suitable soil ; in proof of which we

C

need only leave a basin of paste in their way, and we shall often find it overgrown with "mould" in a single night.

Far lower again in the scale of life than either "mould" or mildew, are the various organisms bearing the general name of "bacteria" (Fig. 8), which are the agents of all putrefaction and fermentation, and the cause of many diseases. These swarm in all moist places of the earth, and are wafted into the air in immense numbers.

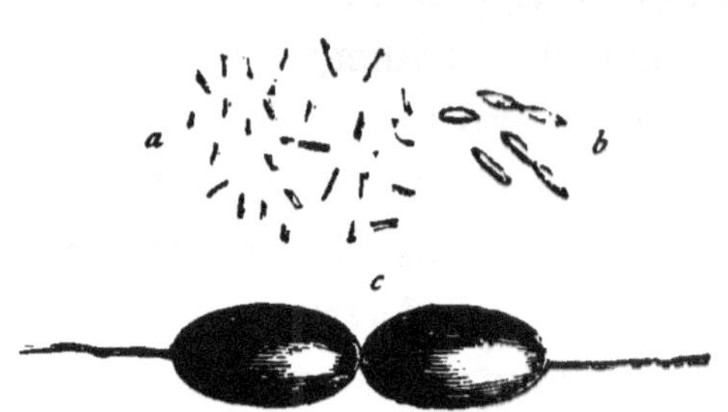

Fig. 8.—BACTERIA: *a*, 600 times their natural size ; *b*, 1,000 times natural size ; *c*, Bacterium, 5,000 times natural size.

Such, then, is the floating matter of the air, and we must now see how its absence would affect us.

We have said that but for the floating dust we should probably have no mist, or fog, or clouds.

Steam, or water-gas, is, like other gases, invisible ; but, as it escapes through the spout of the kettle the cooler air of the room makes it condense into a cloud, consisting of numberless minute globules of water. But the air of the room, besides being cool, is also dusty; and if the steam be received into a glass vessel containing only perfectly-filtered air, no cloud at all will be formed, but the steam will condense into dewdrops upon the glass.

Hence, from repeated experiments, it seems that whenever water-gas condenses into mist or fog, each globule forms round some minute particle of matter, and when our breath becomes visible, it is a proof that the air we are breathing is not only cold, but more or less dusty.

Microscopic meteorites are often found in the centre of hailstones, as if the ice had crystallised upon them; but the invisible dust produced by a gas-jet, a clear fire, the heating of the hundredth part of a grain of iron, or a fragment of glass, is quite enough to cause steam to condense into a cloud; while common salt burnt in a spirit-lamp produces intense fog, as soon as the steam is admitted, and burnt sulphur gives a fog so dense that it is impossible to see through a thickness of even two inches.

As it is calculated that more than two hundred tons of sulphur are burnt daily in London with the coal during winter, this alone, without the suspended soot, would be enough to account for much of our fog. Happily the human furnaces are so constructed as to consume their own smoke, otherwise, since eight hundred tons of carbon as well as nearly 2,000,000 pints of water, are said to be discharged daily from the lungs of Londoners alone, the consequences would be dismal indeed. ·

Still, even when we have got rid of our smoke, we shall not be free from fog, though we shall see no more of the "pea-soup" variety. Fogs, as we know, prevail out at sea, and even high up among the mountains, where there is certainly no smoke, though there is dust; and as long as the air is dusty fogs there will be.

Well, we should most of us be willing enough to dis pense with them ; some of us, indeed, would be more than willing, especially the sailor, who has fog-horns sounding all round him as he goes up and down Channel, and often dares not close his eyes for many days and nights together, and lives in a state of constant anxiety as long as the fog lasts.

But if the air were quite free from dust we should also lose the clouds which make so large a part of the beauty of our skies; the artist would lose his mists and atmospheric effects, and we should probably have no rain, for when the moisture became more than the air could hold, it would be deposited as dew upon every object with which it might come in contact. Neither is this all that the absence of dust would entail upon us.

We might suppose that with no dust in the air we should at least have more light; but while it is undoubtedly true that the sunbeams show us the motes, it is no less true also that the motes and finer dust actually show us the sunbeams, and that one is invisible without the other.

A beam of sunlight or electric light if admitted into a chamber of which the air is perfectly pure at once *disappears*, and is replaced by pitchy blackness, except where it strikes the wall or some other object. Balloonists tell us that the higher they ascend the deeper becomes the colour of the sky, until at the height of a few miles it looks almost like a black canopy, because, though the sun is shining in unclouded splendour, there is little or no dust to scatter his light. The space between the stars— stellar space as it is called—is, accordingly, absolute blackness, notwithstanding the blaze of light which passes through it and becomes visible on striking our dusty atmosphere.

This universal dust is kept out of our lungs, where

it would be injurious, by the innumerable fine hairs or *cilia* which cover the air-passages, and which, as they constantly wave upwards, filter the air very effectually; but when the strain is too great and prolonged, the hairs cease to act, the membrane of the air-passages becomes inflamed, and bronchitis or asthma follows.

The dust of coal-mines and that caused by grinding, especially steel-grinding, and the polishing of pearl-buttons, marble, &c., particularly where emery is used; also the dust in potteries and china-works; the organic dust and fluff of shoddy- and flax-mills; as well as that arising from the sorting of type, are all injurious and some of them fatal in their effects upon the air-passages and lungs, which the hairs are quite unable to protect. A seedsman once complained to Professor Tyndall that his men were made quite ill during the busy season by the irritation produced by the dust from the seeds, and gladly accepted his suggestion that they should be provided with respirators made of cotton-wool tied up in muslin, which filtered the air so perfectly that no further complaints were heard.

The "black lungs" of colliers are well known, and stony dust is found deposited in the lungs of stone-masons, but, under ordinary circumstances, the natural filtering apparatus is quite effectual, the particles being arrested by the hairs above mentioned and then sent back into the air by the expired breath. The air which we breathe out at the end of an expiration is so absolutely free from dust, that, if breathed across the track of an electric beam, the latter will be pierced by an intensely black hole, for the reasons already given.

So much, then, for the dust of the air; but we have yet to say something about the dust of water; and here again, though the amount of impurity may be so infinitesimal as to be hardly capable of expression in numbers, and though the individual particles suspended may utterly elude the microscope, yet their presence here, as in the air, is revealed to us by light.

Good drinking-water appears simply dirty in the electric beam, and even that which has been filtered through charcoal is seen to be thick with fine suspended matter; but by far the greater number of particles are not only invisible to the naked eye, but are beyond the reach of the microscope, and reveal their presence only by colour. In the purest water, obtained, with extra precautions against contamination by the air, from the middle of a block of ice, the electric beam when passed through it appeared of a delicate blue, purer than that of the sky, and therefore produced by particles finer than those suspended in the air. All the evidence, however, points to the conclusion that in perfectly pure water the last trace of colour would disappear, and the beam would be as invisible as it is in perfectly pure air.

The River Rhine flows into Lake Constance muddy with the sediment it has brought from the mountains; but on emerging partially filtered at the other end it is of a dark but transparent green. The water of the Lake of Brientz is also a deep transparent green, but that of the Lake of Thun, which it feeds, being more perfectly filtered, is a clear blue.

Sea-water of a yellow-green Professor Tyndall found

to be thick with suspended matter, in fact muddy; green water was thick with finer particles; but as the water became of a clearer, brighter green these diminished in size and quantity. Water of a cobalt blue was much more pure, of an indigo tinge purer still, and when almost black, with but a trace of indigo, as is the case in the mid-Atlantic, it was nearly absolutely pure.

The purest natural water, like the purest natural air, is never free from some minute quantity of dust; if it were it would be as black as ink, and, though reflecting a glimmer from its surface, as ink does, would lose all the wonderful play of light and colour which makes so much of its charm.

An inky sky, an inky sea, general inkiness, in fact, except in the direct track of the sun's rays—these, it seems, are some of the results which would follow it we could succeed in banishing all dust from the air and from the water.

CHAPTER II.

DUST-MAKERS—FROST, HEAT, AIR, AND WATER.

Dust-makers—How Rocks are affected by Cold, Heat, Air, and Water—
Expansion of Water and Minerals—Dissolving Power of Water—
Formation of Oxides and Carbonates.

THE first of Nature's dust-makers whose manner of working we are going to consider is frost.

If a hollow ball of cast-iron be filled with water, carefully plugged, and then exposed to a temperature low enough to freeze the water, what will be the result?

The water, as it congeals into ice, will require one-fifteenth more space than it did before, and in the effort to obtain it will exert so much force as either to split the iron or drive out the plug. Bombs measuring more than thirteen inches in diameter have been thus burst in two; while in one experiment, an iron plug, weighing more than three pounds, was hurled to a distance of 328 feet.

Of course, if the iron expanded also, and in an equal degree, there would be room enough for the ice ; but, like most other bodies, iron shrinks under the influence of cold, so that there is actually less space instead of more.

We all know what frost can do in the way of cracking a glass in which water has been left, or, worse still, in bursting our water-pipes, though some people seem to imagine that these latter are burst by the thaw, and not by the frost. Like the iron ball, they are, however, cracked by the pressure of the ice, although we, of course, are not made aware of the fact until the ice melts again. If empty when the frost sets in they do not suffer.

But though we may grumble at the plumber's bill, or perhaps at somebody's carelessness, we should be in a yet worse plight if water did not expand on becoming ice, for in that case the greater part of Europe would probably be uninhabitable, and all the fishes in lakes and rivers would die. The water at the surface, as it grew cooler, and, therefore, heavier, would sink to the bottom, while the warmer, lighter water from beneath would take its place, and be cooled in its turn, and again sink, and this exchange would go on until the whole mass of water was congealed into solid ice, which the summer heat would be powerless to melt. As it is, however, no sooner has the whole of the water in a pond been cooled down to 4° C. (39⅓° Fahr.) than this vertical movement ceases; and although the water on the surface may continue to lose heat, instead of shrinking together, and so becoming heavier, as it has done up to this point, it now does precisely the reverse. Below 4° C. it expands, becomes consequently lighter, remains on the surface, and is frozen into a crust of ice.

But how does all this affect the rocks? for they are our present concern.

All rocks are more or less jointed, because all have undergone drying or cooling, and have shrunk somewhat in the process, and while some, like basalt, show regular lines of division, others, such as chalk, clay, gravel, and even sand, show certain irregular lines, along which water, of course, makes its way more readily than in any other direction. But besides admitting water into these cracks, most rocks also absorb more or less, according to their texture and composition. One coarse kind of granite, for instance, has been found to suck up as much as four pounds of water to the cubic foot in the course of eighty-eight hours, when kept entirely submerged; and although rocks are seldom so severely tried as this in Nature, they must take up a certain amount of water in one way or the other, and when this remains sufficiently near the surface to be frozen, every single drop of it expands with such force as to splinter the surface and widen the cracks within, thus affording easier entrance to the rain on the next shower.

The cliffs along the Hudson River are piled up to more than half their own height with the immense heaps of fragments which have been detached in this way, the winters in the State of New York being so severe that the frost penetrates to a great depth below the surface.

High up among the mountains this work of destruction goes on, not in the winter only, but almost daily throughout the year, the scorching heat of noon, which melts the ice and snow, withers the grass, and blisters the face, being often followed by sharp frost at night. And the result of these extreme and constantly-repeated changes is, not only that the mountains are cut and carved into such bold pinnacles as are seen around Mont Blanc, but that many an apparently solid ridge of rock is found, on inspection, to consist of large angular fragments, still in the position they occupied when united, but so loosely piled together that a gust of wind might scatter them.

Even in Great Britain the wreckage produced simply by

changes of temperature is enormous, the tops of our higher hills being often covered to a great depth with their own ruins; while in Kamschatka both hills and table-lands are crowded with the great square blocks and slabs which have been forced off by the mighty energy of the frost. Even the soil is made to feel its power; fence-posts are sometimes lifted out ot their places by the heaving occasioned by the formation of ice in its interstices, and the farmer finds that the clods on his heavy land are more effectually broken up for him by a sharp frost than they could be by any human implements.

But heat also has a share in breaking down the rocks, and we might burst our iron ball just as effectually by boiling as by freezing the water contained in it.

Between $0°$ C. ($32°$ Fahr.) and $4°$ C. ($39\frac{1}{5}°$ Fahr.) water, as we have seen, expands when cooled and contracts when heated; but above and below these temperatures it follows the usual law, and like any other body, whether solid, liquid, or gas, the more it is heated the more room it wants, one cubic inch of water being enough to produce 1,728 cubic inches of steam.

All bodies do not expand in the same proportion, however, and while most metals remain solid, *i.e.*, frozen, even under the fierce sun of the equator, quicksilver is never anything but liquid, even during intense frost, and expands so rapidly under the influence of heat, that thermometers sometimes burst from simple exposure to a hot sun.

Iron, though remaining solid, expands perceptibly, and the difference in length of the 400 miles of rails laid down on the line between London and Edinburgh is 338 yards, according as they are measured in the winter or in the summer. It is for this reason that a space is always left between each length of rail, for, if it had not room to stretch, it would bend upwards or outwards, with disastrous results.

It is, perhaps, not easy to realise the amount of force exerted by a piece of metal in the act of expansion; but it certainly cannot be ignored with impunity, as the builders of an iron foot-bridge in London have had to learn to their cost. They had covered the iron with concrete, and this again with pavement, forgetting that, although these, too, would expand in their degree, the metal would want to expand still more; and the consequence was that when the summer came, both were split by the force with which the iron swelled upwards.

From this it is evident that rocks composed of two or three different minerals are at a special disadvantage with regard to sudden changes of temperature, as one mineral will expand more than its neighbours, and push them slightly out of place—a mere trifle it may seem, and yet, when often repeated, it will be quite enough to loosen them. In a damp climate the mischief is greatly increased by the absorption of moisture, and its conversion into particles of ice; but in the dry air of the Sahara the rocks are splintered into fragments, and reduced to powder merely by the alternations of heat and cold; and in the Peninsula of Sinai—where the sun has scarcely risen before he begins to peel the skin from one's face, though everything may have been covered with hoar-frost during the night—the very flints become so rotten as to fall to pieces at a touch, from the repeated expansion and contraction they have undergone.

But another of Nature's most active dust-makers is the air, which, though we breathe it with impunity, is yet a deadly enemy to many of the rocks.

This air is a mixture, not a compound, consisting chiefly of oxygen and nitrogen, in the proportion of about twenty-one parts of the former to seventy-nine of the latter, with a very small quantity ($\frac{1}{2500}$) of carbonic acid, and a varying amount of water-gas, or aqueous vapour.

Minute quantities of everything capable of assuming the gaseous form are also to be found in the atmosphere, but these four together constitute its main bulk. The oxygen and nitrogen are simple substances, and cannot be split up into anything else, whereas each molecule of water-gas is a compound, consisting of two atoms of hydrogen and one of oxygen ; and each molecule of carbonic acid gas consists of two atoms of oxygen and one of carbon.

Now, oxygen will combine with every known elementary substance but one, and foremost among those by which it is particularly attracted stands iron, with which it unites to form oxide of iron, or what we commonly call "rust." Not an atom of the iron is lost, but the oxygen it has absorbed has changed its colour, added to its weight and bulk, and made it less compact than before.

In perfectly dry air a mass of iron will remain untarnished, at the ordinary temperature, though if reduced to powder it is so vigorously seized upon by the oxygen as to take fire spontaneously and burn away to oxide. But natural air is never free from moisture, and this fact enables the oxygen to attack both iron and steel wherever it finds them. We have only to rub away the rust from a knife to see how the surface has been roughened by the eating away of the steel.

Those who have possessed specimens of a mineral popularly known as "fool's gold," which looks, as to colour, something like tarnished silver, may have been puzzled by finding the cabinet drawer strewn with an ash-like powder, which constantly re-appeared at intervals of only a few weeks. And when at last they have traced it to the "fool's gold" and found that this was gradually crumbling away, perhaps they were hardly less puzzled than before.

In spite of its name, it has nothing to do with gold, being in fact a variety of iron pyrites, called marcasite, which is a compound of iron and sulphur, both of which attract oxygen.

The oxygen first attacks the sulphur, with which, and the moisture of the air, it forms sulphuric acid. This, in its turn, seizes on the iron, and as the water evaporates, minute needle-like crystals of sulphate of iron are formed. Some of the pyrites is converted into this "salt," as it is chemically called, and more is broken up by the crystals as they force their way to the surface.

Zinc is able to resist the attacks of oxygen at all ordinary temperatures, and it is therefore used in thin sheets as a covering for iron, which, when thus protected, is said to be galvanised. As long as the zinc remains entire, the iron is quite safe, but the smallest hole is sufficient to admit the air, and then, as the iron rusts, the zinc is forced up, because, as was said just now, oxide of iron takes up more space than the iron alone.

Now, many rocks contain a quantity of iron, and to them, of course, oxygen is a great enemy. The very compact and almost black rock, called basalt, which forms the columns of the Giant's Causeway, contains much iron, and where

Fig. 9.—ROCK FORMATION, GIANT'S CAUSEWAY, IRELAND.

exposed to the weather its surface is always found covered with soft brown rust, which gradually eats deeper into the mass (Fig. 9).

Moreover, though compact, basalt is very absorbent ;

wet patches are sometimes found in the middle of its columns, and as all water contains dissolved oxygen,* wherever water penetrates, there brown patches of oxide are formed, which crack the rock if they cannot otherwise find enough room.   There are many other minerals besides iron with which oxygen unites to form oxides, but as the results are similar we need not dwell upon them now, and may go on to consider some of Nature's other labourers.

One of the most active of these, even chemically considered, is water, for though we call many substances insoluble, gold and platinum are probably the only two which are actually unaffected even by pure cold water. One sometimes hears people say that china and glass grow thin by repeated washing, and though their senses must be remarkably keen if they are able to detect the change, there is, no doubt, the proverbial grain of truth in what they say; for, if water be kept boiling in a glass vessel for any length of time, it does certainly dissolve some of the glass, while powdered glass at once gives a perceptible flavour even to cold water, and certain kinds of Chinese porcelain have been proved to be more or less soluble.

But we are speaking of pure water, whereas, from the very nature of the case, it is impossible that there should be any such thing as perfectly pure water, either in the

* It is this dissolved oxygen, not that combined with the hydrogen, which fishes breathe. If you remove the combined oxygen, water ceases to be water, but you may expel the dissolved oxygen and other gases by boiling, and the water merely becomes "flat."

air or on the earth. It is almost pure, indeed, when drawn up by the sun from the ocean, but, the moment it falls to the earth again it begins to gather up various impurities, according to the nature of the soil over which it passes; and even before it touches the ground, during its passage through the air, it gathers up, not only dust, but gases, of which the most important to us just now is carbonic acid, or carbon dioxide, as it is called by modern chemists.

The quantity of carbonic acid in the air is relatively small; but there are from twenty to a hundred times as much in the soil; it is poured forth in large quantities from cracks in the earth in volcanic districts, and is almost universal at great depths, so that, from one source and another, all rain, spring, and river water is more or less impregnated with it, and its solvent powers are thereby much increased.*

Iron oxide and various other oxides, which are called insoluble because they may be kept a long time in pure water without any perceptible alteration, are readily dissolved by water containing carbonic acid; but the gas is also quite capable of working on its own account, without the assistance of oxygen. It is the especial enemy of all rocks containing felspar, as this mineral is composed partly of potash or soda, together with lime and magnesia, and with all these carbonic acid readily unites to form carbonates, which are then easily dissolved and washed away by the rain.

---

* Ordinary springs and rivers contain less than one per cent. ; but water from the Saratoga springs contains 231 cubic inches to the gallon.

D

Of the rocks containing felspar, the one with which we are perhaps most familiar is granite (Fig. 10), which consists of the three minerals, felspar, mica, and quartz, crystallised together in very varying proportions, and in crystals of very various sizes.   Quartz is nearly pure silica ;*

Fig. 10.—SECTION OF CORNISH GRANITE, SHOWING FELSPAR, ETC.

it is so hard as to scratch glass, is but little affected by heat or cold, or by the gases of the atmosphere, and absorbs so little moisture that even frost cannot greatly injure it.

Hence a granite which contains a large proportion of quartz is one of the most durable rocks there is. It is the felspar † which admits the enemy ; and when its potash, soda, lime, or magnesia, have been converted into soluble carbonates and washed away, the silicate of alumina, which

* Flint is impure silica.   † See Chap. viii.

is all that remains, is reduced to powder, and the grains of quartz and mica, having lost the cement which united them, fall apart and form silicious sand. Some granites contain so little quartz and mica that they are called felspar rock, and these are very liable to decay, since, besides being attacked by the carbonic acid, they are also so absorbent of water that the winter frosts make great havoc with them.

But it is the various limestones which suffer most from carbonic acid, especially in towns, where the air contains a far larger proportion of the gas than it does in the country.

All limestone, chalk, and marble, properly so called, consist mainly of carbonate of lime,* which is insoluble in pure water ; but when it comes in contact with carbonic acid, each atom of the carbonate will take up a second atom of carbonic acid, and, having done this, will melt as easily as sugar or salt. This we may readily prove by getting from the chemist some lime-water—water in which lime is dissolved. We should not know from its appearance that there was anything in it ; but, on our blowing into it through a tube, it at once becomes milky, because the carbonic acid of our breath has united itself to the lime and made it into the single carbonate of lime, which, being insoluble, cannot be hidden in the water. If, however, we go on blowing, the water will become transparent again, for the lime will take up more carbonic acid, and the double carbonate will be formed, and melt away, making the water " hard."

* A compound of lime and carbonic acid.

Most spring and river water contains some of this dissolved double carbonate; when boiled, half the carbonic acid is driven off, and the single, insoluble carbonate is deposited in kettles or boilers, forming the crust which we call "fur."

Boiled water is "flat," however, from the loss of the dissolved gases, and it may be as effectually softened in another way. The Canterbury water, being derived from the chalk, naturally contains a great deal of the dissolved bi-carbonate, or double carbonate, and though very clear is very hard. It is softened by the addition of lime (calcium oxide), each atom of which takes away one atom of carbonic acid from the dissolved bi-carbonate, thus converting both the latter and itself into the insoluble single carbonate, which makes the water milky at first, but gradually sinks to the bottom and leaves it quite clear.

But enough has been said to show the strong attraction existing between lime and carbonic acid; and there is no difficulty in understanding how it is that the Portland stone of which St. Paul's Cathedral was built not two hundred years ago is already beginning to moulder, and how tombstones of Italian marble lose their polish in a year or two, and in the course of sixteen years are often so roughened that the grains of carbonate of lime may be rubbed off with the finger.

From these examples we may also form some idea of the way in which carbonic acid works away at the rocks; but on this head there will be more to say hereafter.

CHAPTER III.

DUST-MAKERS—WIND, WAVES, RAIN.

The Sand-blast, how it Cuts and Polishes—How the Waves Work—A Shower of Rain, what causes it, what becomes of it—Springs, Land-slips, "Stone Rivers."

OF late years a new way of grinding and etching glass has been invented by an American. Instead of the usual acid he employs fine quartz sand, which is driven against the glass by a blast of air, and in from ten to fifteen seconds removes all the polish from the surface.

Window-glass exposed near the sea-shore soon loses its polish in like manner from the natural sand-blast which so often drives against it; but the artificial sand-blast not merely de-polishes but actually cuts through substances harder than the sand itself. A solid block of corundum, for instance, which is but little inferior to the diamond in point of hardness, may be bored to the depth of an inch and a half in five-and-twenty minutes; while a quarter of an inch of hard steel may be cut through in ten minutes, and in the course of an hour a piece of marble half an inch thick can be covered with an elaborate open-work pattern, such as it would take many days to produce by the ordinary process.

And now let us see what Nature has accomplished by means of her sand-blast, which has been in operation for ages past (Fig. 11).

In the neighbourhood of Suez the ground is thickly

covered with red sand, which is so fine as to be set in motion by the slightest breath of air.   Even on a still day it looks as if a thin cloud of smoke were constantly sweeping over the surface, while if the wind be but moderately strong, it

Fig. 11.—SAND GLACIER OVERWHELMING A GARDEN IN ELBOW BAY, BERMUDAS. *

is almost impossible to face it, owing to the cutting sharpness of the sand with which it is laden ; and the hands when held close to the ground at such times, tingle unbearably for the same reason.

* Sir Wyville Thomson says that the sand has entirely filled up a valley, and is steadily progressing inland in a mass about twenty-five feet thick. On its path from the beach it has covered a wood of cedars !

Nature's sand-blast is indeed a powerful engine, for it has pitted and polished the upper surface of every rock and stone in the desert, besides so smoothing and rounding many a boulder that it looks as if it had been rolled about in the sea.

In other parts of the world a glance at the rocks on the coast is often enough to tell one the prevailing direction of the wind, so much have they suffered on that side. At Heard Island the sand-blast has cut the rocks into tree-like shapes; in the Bermudas (Fig. 11) it has given them an artificial appearance as of being cut and dressed by a mason's chisel; in a pass to the south of the Great Salt Lake the hills are worn away and the rocks polished by the ceaseless scouring of the sand driven against them from the west; and in an island in Lake Kara-Kul, not only has the surface of the hard sandstone suffered considerably, but some of the rocks have been drilled through by the sand-laden wind from the north.

Then again, stones have been brought from Lyell's Bay, near Wellington, New Zealand, many of which the uninitiated would certainly imagine to have been shaped by human hands, so strongly do they resemble the knives, arrow-heads, spear-heads, &c., of what is called the Stone Age. All have sharp-cutting edges, and their facets seem to have been chiselled with careful attention to symmetry. Yet the only chisel used upon them has been the sand of the bay, and the hands which guided it were the two winds there dominant, which have urged the sand against opposite sides of the stone in turn, each grain of sand chipping away its infinitesimal bit of stone, and in the end sculpturing these singular forms.

So much for the work done by wind-driven sand. We shall see presently that it is no less effectual when urged by water.

The waves of the sea, indeed, as they thunder against the cliffs, are powerful enough to do a large amount of rough work without assistance; but what tremendous blows they must deal when they rush up to a height of perhaps a hundred feet, laden, as of course they often are, not only with sand and pebbles, but even with large stones.

When especially furious, they will tear down piers and jetties and carry away long lengths of sea-wall, and at all times they are more or less busy undermining the cliffs, and when these at last fall in by their own weight, the fragments are dashed one against the other, broken into smaller fragments, rolled and rounded into pebbles and ground into sand, and perhaps swept farther along the coast and cast up as beds of sand, gravel, or shingle.

With all their fuss and fury, however, the waves are usually unable to keep pace with the silent workers above —frost, thaw, atmosphere, &c. And, accordingly, though the cliffs are hollowed into caverns here and there, they do not on the whole overhang the sea, but slope away from it, showing that the wear and tear proceed after all at a more rapid rate above than below.

Perhaps the largest shingle-bed in the world is that which lies on the east coast of Patagonia, and is between six and seven hundred miles in length. Its average width is two hundred miles, and its average thickness fifty feet, which, in at least one place, is increased to more than two hundred feet; and if the pebbles alone, without the accompanying sand and mud, were piled together, they would make a great mountain chain.

Running water, when provided with a due amount of sand or gravel, is capable of undermining and rounding the hardest rocks, as we shall see more particularly when we come to speak of the rivers, but we have first to say something of their mother, the rain.

What happens after an ordinary shower?

For a time the surface is wet and there may be puddles in the hollows, but if the soil be a light one, all traces of the rain soon disappear, and we say it has "dried up." Every drop that has fallen *is* somewhere, however, and it has certainly not all gone back to the air in the form of vapour, though in hot or windy weather so much may do so that the soil will benefit but little.

The hotter the air the more water it can take up and hold in the form of invisible vapour, but it can never hold more than a certain quantity at any given temperature, and when it has taken up this utmost quantity it is said to be saturated. In our damp island the air is often saturated with moisture, which is not an agreeable state of things; but more unpleasant still is it when the air contains less than half the saturating quantity, for it will suck up moisture wherever it finds it, and therefore takes it, not only from the ground but from vegetation, and even our bodies, making us feel "dried up" and wretched. In the African deserts, where the very air, as well as the ground, is parched with thirst, the traveller finds that his lips crack and bleed, his whole body burns, and his skin is dried till it bursts in a hundred places. Even then, however, the air is not absolutely dry, though it contains perhaps not more than a fifteenth of the saturating quantity. If it were quite dry, no animal or vegetable could exist in it.

But besides preserving all living things from being dried to chips, the moisture of the air also benefits us in another way, by keeping the earth both cool and warm. By day it acts as a screen to moderate the fierce heat of the sun, and by night it serves as a blanket to keep in some of the heat received during the day.; and accordingly, where the air is comparatively dry, as in the desert, the nights are intensely cold while the days are scorching hot.

When warm air laden with moisture rises into higher and colder regions, or meets a current of colder air, some of the moisture becomes liquid and condenses into a cloud of very minute globules. This is what happens in the case of our breath on a cold day; and though whales have been popularly believed to suck in the jets of water which they throw up, they, too, really discharge only breath, the moisture of which condenses and falls down like a fountain from a very fine rose, or a shower of rain from a cloud.

And this brings us back to the shower, whose history we are going to investigate. Part of it, then, has returned to the air, part has been sucked up by the roots of trees, &c., and the remainder sinks farther into the ground and goes on sinking until it meets with something to stop it. In a very light, porous soil, the water drains through so rapidly that none but the least thirsty plants get enough to satisfy them, and the ground is consequently barren.

If, on the other hand, the soil be close and impervious, like clay, very little water will sink in, and the greater part will either stand in ruts and puddles until sucked up by the air, or run away to feed the nearest ditch, or sink through some bed of sand at a distance. But though clay usually lets but little water pass through, yet it parts with so much of its own moisture in hot dry seasons that it gapes in all directions, and when the rain comes much of it escapes through these cracks.

In one way or other, then, a great deal of rain sinks into the ground, and when its downward progress is arrested by some impenetrable bed, it soaks through the soil sideways, and sometimes travels many miles before it finds an outlet, while, by dint of often using the same road, it frequently wears for itself an open channel, along which it flows as rapidly as the rivers above ground do in theirs. That this must be the case is evident from the fact that a large increase of water is observed in certain wells and fountains (as, for instance, at Nismes), so soon after rain has fallen some miles off that it is impossible it should have passed through soil, however porous. Moreover, in some places the water from time to time throws up seeds, vegetables, fresh-water shells, and even fishes, which have evidently travelled a long way underground, as much sometimes as one hundred and fifty miles, and this, of course, can only be by open channels.

Where such channels do not exist the water spreads about in all directions in search of an outlet, which it finds, perhaps, in the face of a cliff, the side of a hill, or at the bottom of a slope so trifling as not to deserve the name of hill. Where, however, the whole face of the country is as flat as a table, there, of course, no springs can come to the surface, and the water remains locked up in its subterranean reservoirs; and where there are no springs there are neither brooks, nor streams, nor rivers, and the inhabitants must either sink wells, or catch and store the rain as it falls.

But it is as dust-makers that we must now consider the springs, and in this capacity they are often very active.

As they flow underground they tear up and carry away in the form of mud some of the clay, &c., over which they pass, thus undermining the rocks above, which in course of time sink down or subside. But if the bed over which the water flows be, as it often is, a sloping one, something else is very apt to occur; for if the upper beds be of sand, chalk, or any other porous rock, they become so heavy after much rain that large masses often slip down over the greasy clay beneath, which, instead of offering any resistance, helps them down just as the tallowed planks help to launch a ship.

Landslips of this kind have occurred at various places along the English coast, and not long ago one took place at Crich Cliff, near Matlock. The limestone of which the hill consists rests upon a sloping bed of soft clay, which was rendered softer and more slippery still by unusually heavy rain. Cracks opened in the hill-side and in some cases grew from a few inches to yards in width. Soon after, the cliff was seen to be in motion, and great masses of lime-stone, many tons in weight, dashed down the hill, with a roar like thunder, and swept away the trees in their path as if they had been so much grass. Many million tons of rock were carried down in this way; but though a landslip of this kind is alarming enough, it is a mere trifle compared with what takes place in other parts of the world, as, for instance, in Switzerland, where the Diablerets, a mountain which is limestone above and soft shale below, has lost three out of its five peaks, and almost filled a valley with its ruins.

In many mountainous districts landslips are of constant occurrence in the wet season, and the sides of the White Mountain (New Hampshire, U.S.A.) are deeply furrowed and scarred by the masses of earth, &c., which have been hurled down them. In this case, while part of the mountain itself is sometimes carried away, the landslip more often consists of the loose upper surface, the cap.of soil, in fact, with the forest-trees growing in it, which is stripped off to a depth of fifteen, twenty, or thirty feet, leaving the rock quite bare over an area sometimes ot many acres.

Another curious form of landslip is that of the famous "stone-rivers" of East Island, one of the Falklands, which are produced, not by any extraordinary convulsion, but by the wear and tear of every day.

These stone-rivers at a distance much resemble glaciers, and vary in width from a few hundred yards to a mile or two, while the irregular blocks of stone of which they are composed are from two to twenty feet long. They are derived from the bands of quartzite of the ridges above, some of which are very hard, and others so soft that they pass into crumbling sandstone. As the latter are worn and washed away by rain, frost, &c., the compact bands are left as long projecting ridges along the crests and flanks of the hill-ranges, until at last, deprived of support, they give way at the joints, and fall from their places; then, when once they are embedded in the soil, the whole mass creeps down even the gentlest slope, and so fills the valleys.

But rain does more than produce springs and landslips. A heavy fall will do a certain amount of direct damage, even in this temperate climate, by washing away gravel, and making watercourses down the sides of the hills and along the roads; but in the tropics its operations are on a much larger scale. There it comes down, 'not in drops, but in strings, and is often so heavy that fresh water may be scooped up from the surface of the sea!

After seven hours' rain in Brazil, on one occasion, there were torrents rushing down every slope; the water streamed through the roofs and out at the doors, and it was feared that the whole village of Capelhinha would be washed away. At such times the rain descends in streams so dense that the level ground is quite unable to absorb it, and is quickly covered with a sheet of water, while it rushes down the hills in torrents of such volume as to wear deep channels in their sides, and to wash away all the mould from the rocks.

A few miles from Sydney, where the soil is friable, a chasm, twenty feet deep and twenty yards across, has been scooped out by the rain in the course of a dozen years or so; and the Blue Mountains of Australia have been cut into extraordinary deep gullies and chasms by the same agent.

It is not often possible to distinguish between the effect of driving rain and that of running water, as the two are generally combined; but in several places, notably in the Ravine of Finsterbach, columns of hardened mud, from twenty to a hundred feet in height, have been cut out, and separated from the terrace of which they formed part, by the action of rain alone.

A similar phenomenon on a larger scale may be seen among the Wasatch Mountains of the Western States of America, where pinnacles, some of them four- hundred

Fig. 12.—BLOCK OF PLUM-PUDDING STONE.

eet high, fringe the bank of the South river for miles. The plum-pudding stone (Fig. 12) of which they are composed (rounded pebbles cemented together by clay and sandstone) is liable to crack in dry weather, and the rain, eating its way down through these cracks, wears long grooves in the softer stone which it meets with below. A cap of harder stone usually remains for a while on the top of the pinnacle, and protects it from the weather, but when this is blown or worn away the whole monument crumbles down by degrees. A thickness of four hundred feet, and some square miles in extent, of solid rock is entirely gone, with the exception of these pinnacles ; but though worn down by the effect of rain and weather, it has been removed by running water.

CHAPTER IV.

DUST-MAKERS AND DUST-CARRIERS—RUNNING WATER.

Running Water: the Tools with which it Works—The Colorado and its Cañons—Loads, seen and unseen, transported by Rivers—Caverns in Limestone Rocks.

SOMETHING more than three-fifths of the rain which falls over England and Wales sinks into the ground and goes to feed the springs, from which, in dry weather, except in the neighbourhood of snow-clad mountains, all brooks, streams, and rivers, derive their whole supply of water. And as these increase in volume the farther they flow, it is evident that they must receive supplies, not only at their source, but at various points along their course. The river Churn, for instance, starts with a flow of eleven cubic feet of water per minute; but a quarter of a mile from the spring head, though it has not been joined by any visible tributary streams, the flow has increased to thirty-one cubic feet, and at a distance of five miles and a half it has increased again to 320 cubic feet, so that other and perhaps many small springs must have discharged themselves into it.

Now all running water, whether it flows above or below ground, has some power of wearing away its channel. Flowing at the rate of but three inches a second, it will tear up fine clay; at double this speed it will remove fine sand; at twelve inches it will sweep away fine gravel, and at three feet it will roll along stones as big as an egg ; while mountain torrents which have dwindled to mere threads, swell in a few hours to such a size that they will carry before them sand, mud, rocks, and trees, will sweep away bridges, and bury meadows ten or fifteen feet deep in rubbish.

Then, again, a mere dribble of water flowing perpetually over limestone rocks will, in time, produce results such as seem to be quite out of proportion to its size ; and we may, in fact, safely affirm that even the most sluggish stream wears its bed more or less, in one way or another, either chemically or mechanically.

It is when they are provided with tools of just the right sort, however, that streams and rivers get through most work in the way of carving and grinding.

From the banks, cliffs, &c., between which they flow, they receive constant supplies of mineral matter, which, having been loosened by alternate frost and thaw, are washed and blown into them by rain and wind. The more rapid the stream, the more quickly these contributions are carried along and the harder blows they give to the sides and bed of the stream, as well as to one another. The larger fragments are rolled along the bottom, deepening the bed, wearing off one another's sharp corners, rounding the angles of any rocks they may meet, undermining the cliffs and pounding the smaller fragments like so many pestles in a mortar, until they are reduced to gravel, sand, or mud, which is then floated farther down the stream.

It is by the gravel and sand, though they look so much less important than the boulders, that the river's chief work is done ; but much depends upon the inclination of the bed

E

and consequent rapidity of the current, as well as upon the nature of the rocks with which it has to deal.

Even a small rivulet when flooded will transport from one to three thousand tons of gravel in a day; and during the rainy season in India, when every river and mountain torrent are swollen, when for days and nights together nothing is heard but the crash of trees and boulders, and great masses of earth and rock, three or four thousand feet in length, fall from the mountain side and are ground to sand and mud in the boiling waters, the channel of every stream, great and small, is enlarged, and enormous quantities of mineral matter are carried down to the ocean. One river in Bengal removes a depth of ninety feet of stone and earth from its bed every year, and the Ganges brings down in the same time more than enough solid material to build up forty-two Great Pyramids.

The Americans have a saying that in time of flood the "Big Muddy," as they call the Missouri, is "too thick to swim in and too thin to walk upon." For winter in the North-western States is long and severe; the rivers, streams, and brooks are all fast frozen, and the country is covered with colossal heaps of snow, which sometimes attain well-nigh fabulous dimensions. Then suddenly, almost without warning, comes the intense heat of summer; the streams are set free, the snow is melted with lightning rapidity, violent torrents of rain fall, the brooks are swollen into rivers, the rivers into boiling mud-laden floods, which at once burst their bounds, inundate the valleys, overthrow trees, houses, mills, and sweep away whole tracts of land in one place, while they pile up immense islands and sand-banks in another. Owing to the character of the country through which they flow there is indeed little or nothing to restrain their progress; for the high lands are either not wooded at all, or so sparsely covered with trees, that the river banks have but little power of resistance, and the clayey, sandy soil of the prairies, intersected by numerous deep rain-made watercourses, is still less capable of opposing the flood of waters.

It is when we look farther west still, however, that we are perhaps most impressed by the magnitude of the work accomplished by rivers; for here they have to deal not with yielding banks of sand, but with the solid rock, which becomes a permanent monument of their mighty power.

The Colorado, with the Green River, as it is called during the first part of its course, is 2,000 miles long, and the upper two-thirds of the basin which it drains lie at a height of from 4,000 to 8,000 feet above the sea, surrounded by snow-clad mountains. During winter the snow falls heavily, filling all the gorges and covering all the hills; and when summer comes and the great piles melt, millions of cascades leap down from the rocks on all sides. " Ten million cascades," writes Mr. J. W. Powell, " rush together and form ten thousand torrent-creeks; these again unite to form a hundred rivers beset with cataracts, and a hundred roaring rivers unite to form the Colorado, which rolls, a mad, turbid stream, into the Gulf of California."

The current is at all times strong and rapid,* and is provided with exactly the right tools, in the shape of mud,

---

* The velocity of the water and stone in the Cataract Cañon is equal to that of a railway train going forty miles an hour.

Fig. 13.—THE CATARACT CAÑON, COLORADO.

sand, and boulders, the last of which, indeed, while wearing the bed of the river, are still more worn themselves, being finally reduced to mud, which is more or less dissolved before reaching its journey's end. But the sand is the most important tool, and the river has used it to such purpose that it has cut the rock into deep gorges or cañons (Fig. 13), which extènd throughout more than 1,000 miles of its course, and are from 600 feet to more than a mile deep, though in some parts only twenty or thirty feet wide. The great river, when viewed from above, dwindles to a silver thread at the bottom of these gloomy sunless gorges, through whose long length no one in these days is known to have passed alive except the members of the exploring expeditions sent out in 1869 and 1872, though horrible tales are current of the sufferings of those who, having once taken refuge within them, have found escape well-nigh impossible. In addition to the enormous height of the cliffs, there is the further difficulty that they are for the most part undermined, the weathering above being quite unable to keep pace with the rapid working of the river below. The only practicable points of exit, therefore, are at the rare openings made in these giant walls, where the river is joined by its tributaries.

Each tributary of the Colorado, every branch of each tributary, and each little stream and rill, has cut for itself similar cañons on a larger or smaller scale, and hence the whole country is such a perfect labyrinth of chasms, that it is a difficult matter to choose routes for railroads and other traffic.

In the adjoining state of New Mexico, similar cañons, some of them 1,000 feet deep have been cut in the sandstone by the rivers Mora and Canadian. At the junction of the two rivers the cañon was at one time 860 feet deep, but was afterwards filled to a depth of 470 feet by a stream of basaltic lava from a neighbouring volcano. Since those days, however, the united rivers have worked to such purpose that the whole of the basalt and 230 feet more of the sandstone have been cut through, making the gorge now 1,090 feet deep. As a rule, rivers are more apt to deepen than to widen their beds, the friction being greater at the bottom than at the sides ; and usually the " weathering " of the cliffs or banks, produced by frost, thaw, wind, rain, or air, goes on more rapidly than the grinding below. But cañon-making rivers work at greater speed than these atmospheric influences, and the Canadian and Mora, after cutting a narrow passage through the middle of the lava, went on to undermine it, so that it now projects in a wide terrace on either side, deep down within the gorge.

The neighbouring river of La Platte, on the other hand, though its bed slopes on the whole as much as that of the Colorado, and it has plenty of sand, makes no such cañons, for its load is just as much as it can carry, and more than it can use with effect ; its banks, too, are not hard and solid enough to stand as walls, and the loose sandstone of which they are composed is blown or washed into the river as fast as it crumbles down, and almost chokes the stream. Too little sand, on the other hand, will be swept onwards without making any impression, and it is when the quantity is nicely proportioned to the force and volume of the current that the most striking results are achieved.

When pieces of rock are first broken off, whether by the

force of the stream itself or the action of frost, &c., they are, of course, all angles, and as these prevent their rolling along easily, they are carried but a little way at first, and then left to be ground down into more manageable shapes and sizes. Smaller, rounder pebbles are carried farther, and in the case of short rapid rivers are not dropped until they reach the sea ; and rounded pieces of porous pumice-stone are found floating down the Amazons, one and two thousand miles away from the volcanoes Cotopaxi, &c., whence they must certainly have come, although the Brazilians, who use them to remove rust from their guns, firmly believe them to be solidified river foam.

Having travelled so far, they would be pretty sure of being floated out to sea, but their case is exceptional, and, generally speaking, the nearer we come to the mouth of a river the finer is the mineral matter which it carries, and, before it has finished its course, its load has been so thoroughly sorted by the dropping of all the heavier portions, one after the other, that at last it carries nothing but the finest mud, which, on reaching the sea, is then carried farther by waves and currents, until finally deposited in the largest department of the world's great lumber room.

If, however, the river should fall into an almost tideless sea, such as the Mediterranean, Carribean, or Gulf of Mexico, much of the mud, instead of being swept away, is deposited at its mouth ; and this is the origin of the tongue of land at the mouth of the Mississippi, and also of the great triangular plain of Lower Egypt called the Delta, which, like all deltas, has been won from the sea. The Nile has cut for itself in the rocky surface of the desert a great trench, 300 feet deep and from less than a mile to eight miles wide, and the whole of this solid stone it has ground into mud. On an average it brings down nearly 131 cubic feet of sediment every second, so that it would take about nineteen seconds to fill a room seventeen feet long, fifteen wide, and ten high. Much of this sediment consists of the rich soil washed down from the highlands of Abyssinia, and to it Egypt owes her wonderful fertility; for year by year it is spread over the whole surface covered by the inundation, renewing the land, which never need lie fallow nor have any artificial aid to render it fertile, though the rate of increase is only about four inches and a half in a century.

The Mississippi transports a larger load than the Nile, 147 cubic feet per second; while the Ganges outdoes them both, carrying as much as 557 cubic feet.

Compared with these large rivers, the Thames seems but an infant, yet in the course of a twelvemonth it manages to carry down the very respectable load of 1,865,900 cubic feet of solid matter, and the Danube, Po, and Rhone, convey several hundred times as much.

Since the greater part of the mineral matter carried away by rivers is ground extremely fine, as we have seen, it is quite evident that most of it must be conveyed into the ocean, for a specimen of Rhine water, though kept perfectly still, has been found to take four months to become quite clear, and the sediment could not therefore by any possibility have time to settle while being carried down by the river in the few days that would elapse before it reached the sea.

We find, moreover, as a matter of fact, that, while masses of rocks miles in thickness and thousands of miles in extent—whole continents, indeed—have been formed in the sea, the fresh-water deposits of rivers, lakes, and estuaries, are to be reckoned only by some thousands of feet.

The Rhine, as we have seen, drops part of its load in the comparatively still waters of Lake Constance, but though it emerges thence no longer muddy, its dark green colour shows that it is by no means perfectly filtered, and the finer particles are carried farther still.

But besides the mud and those still finer particles which give to water its green or blue tint, rivers carry away a vast amount of mineral matter, which, though absolutely invisible to us, is none the less important. A very small pinch of powdered chalk will, we know, make a large glass of water quite milky, while a handful of salt will disappear, leaving it just as clear as before. The chalk is held in suspension, and in time will settle at the bottom of the glass, while the salt, being dissolved, or held in solution, will not reappear until the water is removed, which may be done either by leaving it to evaporate or by boiling. When sea-water is boiled its various salts are left behind in the form of crystals, and the steam arising from it, if caught and condensed into water, will be found to be almost pure. In like manner the sun draws up large quantities of almost pure water from the ocean, leaving the salts behind.

It is very evident that rivers are most heavily laden with sediment either during the rainy season, when large quantities of mineral matter, loosened by various agents, are washed down from their banks, or when the snow is melting on the mountains and rushing down in numberless torrents to the valleys. It is when the snows melt that the Rhine is at its fullest and muddiest, and often causes the waters of Lake Constance to rise as much as a foot in twenty-four hours, and the mineral which it chiefly conveys is carbonate of lime, which constitutes more than a third of the deposit left in the lake. On leaving the lake, the Rhine again passes through miles of limestone, as do also its tributaries, and yet the sediment which it finally takes to the ocean contains little or none. It does carry it, indeed—*that* we may be sure of—and in considerable quantities, too, but not as visible sediment, for it has been dissolved by the carbonic acid of the water and thus rendered invisible

Up to the time when it reaches the lake, the course of the river is so rapid and tumultuous that the carbonic acid has but little chance of doing anything, and the powdered limestone is simply carried down as sediment thus far. On leaving the lake, however, the river has a long course before it, and when it reaches the sea all the carbonate of lime which itself and its various tributaries have collected, and even all that is washed into it in time of flood, has been so completely dissolved that we should not be aware of its presence but for the hardness it has given to the water.

Next to carbonate of lime, the mineral which is carried down to the sea in the largest quantity is sulphate of lime, which never fails entirely except in the case of a few small rivers. But we shall gain a better idea of the amount of mineral matter removed by the rivers when we come to see what is done with it ; meanwhile we may mention, for those who like statistics, that such rivers as the Danube, Rhine,

Rhone, and Elbe, would, in the course of 8,000 years, have conveyed away in solution an amount of mineral matter equal in weight to the whole quantity of water discharged by them in a year.

In India the various dissolved carbonates contained in the river-water are a source of some perplexity to the cultivators of the soil; for as rain falls only at certain seasons, and there are often long periods of drought, large sums have been expended in conveying water to the fields by means of a system of irrigation, and now that this has been done, it is found that the land in some parts has been rendered actually unfit for cultivation by the large amount of mineral matter conveyed in the water, and the question is how to get rid of it again.

When considering the dissolved minerals carried away by water, we must bear in mind that they are not, like the sediment, taken only, or even chiefly, from the bed of the river and the banks between which it flows. On the contrary, they are drawn from far and wide, from the whole area, in fact, which the river drains. For the rain, as it soaks through the earth, dissolves something of every bed through which it passes * ; the springs, as they flow under ground, often by very complicated channels, do the same ; and, in fact, the whole network of streams, large and small, above or below ground, by which every river of any size is fed, all contribute their share of dissolved mineral matter, and are most strongly impregnated just when their waters are most

* About 100 tons of mineral matter are said to be annually dissolved per square mile, all over the world ; of this about half is carbonate of lime.

transparently clear, that is in time of drought, when the rivers are fed almost entirely by springs alone.

The Thames, with its estuary, receives the drainings of 10,000 square miles; the Severn, of 8,580; the Colorado of some 300,000; and the mighty Amazons, the " Mediterranean of the West," of 2,048,000.   It is because it is taken from such wide areas that the large amount of rock annually removed and carried off to the sea makes so little apparent difference.   Thus, more than 8,000,000 tons are invisibly removed from England and Wales alone each year, and if this were taken equally from every part of the surface, it would be nearly thirteen years before a foot in depth was carried away.   Slowly, but surely, however, all land traversed by streams, of whatever size, is being worn down and conveyed to the ocean ; and where the rocks are of chalk or limestone, the work done is often perceptible enough.

The rivers of the Teutoburger Wald and Haar, for instance, annually take away more carbonate of lime than would make a cube measuring 100 feet each way ; and in sixty-seven days the Pader springs carry off enough to build a cone 150 feet in diameter and twenty-four feet deep ; one result of which is that landslips and subsidences are of constant occurrence in their vicinity.

Scattered about on the table-lands of Wiltshire and Dorset are accumulations of flints, sometimes several feet thick, which once formed beds separated one from the other by many feet of chalk, which has long since been dissolved and carried away; and in many parts of the world vast caverns have been hollowed out in the rocks by the agency of water and carbonic acid alone.

In Styria there is a wild desolate region, where the rocks are so porous that every drop of rain at once passes through them, and the surface is so dry that hardly any green thing will grow. Down below, however, the scene is one of great beauty, for here are the famous Adelsberg caverns, halls excavated in the limestone, some of which are more than 250 feet long, and lofty in proportion, their richly-sculptured roofs being supported by elaborately carved pillars, while many are adorned with statues, obelisks, clustered columns, birds, beasts, trees, plants, &c., all apparently chiselled out of pure white marble, though the only tools used have been water and carbonic acid.

These two have dissolved the rock as they passed through it, and then evaporating, have deposited the carbonate of lime again in these various forms, sometimes as stalactites of every size and shape, which hang from the roof like icicles, sometimes as cement, joining together broken fragments of rock ; sometimes, falling to the ground, they have built up wonderful stalagmitic columns, which vary from a few inches to several feet in diameter, and at others they have covered the walls with what look like festoons of drapery, while in one place they have woven a curtain about ten feet long and only an inch thick, which hangs in the most graceful folds and seems to wave gently to and fro, as the light from the guide's lamp falls on it from above.

All these various forms of ornament are due to the chemical action of water charged with carbonic acid, by means of which some of the carbonate of lime removed from above has here been re-deposited ; but the long, lofty caverns, which extend for miles, have no doubt been excavated by the mechanical action of the Poik, which at one part of its course disappears through an opening in the earth, and flowing underground for several miles, passes through one of the great halls just described on its way.

In South Australia there is a series of limestone caverns, one of which has the appearance of an immense Gothic cathedral, the roof being apparently supported by one huge stalactite, tinted with almost every shade of colour, while the area is occupied by numerous half-finished stalagmites, which look, in the dim light, like kneeling worshippers. The stalactites in each cavern seem to possess a distinct character of their own, and differ one from the other as much as do the leaves of the forest.

In one of the Bermuda caves there are grand stalagmitic columns reaching from floor to roof, one of which is beautifully fluted and fretted with stalactites, and measures sixty feet in circumference; in another the stalactites hanging from the roof are perfectly white, some of them as fine as knitting-needles, and often yards long; while, wherever there is a continuous crack in the roof, there descends from it a graceful, soft-looking, white curtain.

Of all the limestone caves hitherto discovered, however, the most extensive is the Mammoth Cave, of Kentucky, which, with its 226 avenues branching out from the main gallery, is computed to have a total length of about 160 miles.

But our chief point now is, that, whether excavated by chemical or mechanical means, or by both, the vast quantities of limestone which once filled such caverns have been carried away to the ocean by the springs, streams, and rivers, which permeate and overspread the earth.

CHAPTER V.

DUST-MAKERS AND DUST-CARRIERS—GLACIERS AND ICEBERGS.

Floating Ice—Glaciers : the Loads they Carry, their Tools, how they Cut, Scratch, and Polish—Stone-meal—Fjords Engraved by Glaciers— Glaciers in England—Glacial Dust-heaps—When the Great Thaw came—Icebergs as Dust-carriers—The great Distances they Travel.

RUNNING water, as we have seen, can do much in the way of wearing down rocks, and transporting heavy loads ; but frozen water can do still more, and a river filled with miniature icebergs, or broken sheets of ice, possesses tremendous powers of destruction.

When the ice broke up on the Danube a few springs ago, such great masses were hurried along by the tumultuous, swollen waters, and were hurled with such terrific force against anything that came in their way, that the closed gates of a canal were burst open, and a solid wall of masonry, seven feet thick and 250 feet long, was speedily and entirely overthrown.

Ice being lighter than water, is easily borne along, even by a feeble current, and although when loaded with sand, pebbles, and bits of rock, it may be too heavy to be carried on the surface, it will still float down the stream without difficulty.

When the ice breaks up on the St. Lawrence, the huge

slabs pile themselves one on the other, until a pack many feet high is formed, which not only forces along great boulders and blocks of stone some tons in weight, but also breaks huge fragments from thirty to fifty feet square from the cliffs, wharves, and stone buildings between which it passes.

It is not often that the Thames is even nearly frozen over; but a few winters ago it was for some days full of broken ice, among which were fragments covered on the under side with gravel, as if they had been formed in the bed of the river, and had then floated to the surface. In many of the Siberian rivers large stones are very commonly found thus embedded, and are carried down by the ice in considerable quantities.

We were explaining, a chapter or two back, how it was that ice was formed, not at the bottom, but on the surface of water; but ice cannot easily form on a running stream, owing to the constant motion, which is greatest on the surface and in the centre of the current. Then, too, the whole body of water, being kept in constant agitation, becomes so thoroughly mixed that the temperature is the same throughout, and when it falls to 0° C. (32° Fahr.), the comparative stillness in the bed of the river, and the contact with the cold surface of rocks and pebbles, enable the ice to form there, and in the warm sun it floats to the surface with whatever may be adhering to it. In the small tributaries of the Thames, vast numbers of pebbles, and even stones a foot in diameter, are carried away by ground-ice, and other rivers transport still larger quantities in this way.

If all ice formed thus at the bottom of the water, our

lakes and rivers would be perpetually frozen, as has been said ; and on the other hand, without those rivers of solid ice, which we call glaciers, there would in time be neither lakes nor rivers, nor even seas and oceans.

The hot air of the tropics is constantly drinking up many millions of tons of water, and if none were restored to it

Fig. 14.—SNOW CRYSTALS.

again, the whole surface of the ocean would be lowered about eight or ten feet every year, and in time must be utterly exhausted. But the water thus taken up and transported thousands of miles by the currents of the air condenses into rain clouds or snow clouds, according as the temperature which it encounters is above or below freezing point, and after a longer or shorter journey, is poured back into the ocean by the rivers.

Snow consists of crystals of ice (Fig. 14), which look

F

white* only because they do not lie perfectly close ; for when the air is squeezed out they adhere together and form a lump of transparent ice. The snow which collects on the heel of one's boot is converted into ice by pressure ; and the vast quantities of snow which fall and accumulate among the mountains are similarly converted into ice by their own weight, which also causes them to creep slowly down the mountain sides into the valleys, where the warmer air changes them once more into water.   If the glaciers, as these rivers of ice are called, remained stationary high up among the mountains, they would go on increasing in thickness year by year, as they received fresh additions of snow, and year by year, as its waters were locked up in the form of ice, instead of being returned to it, the ocean would sink lower and lower.

Glaciers, then, may be called rivers of ice, but unlike other rivers, they are able to move uphill as well as down, and while at one time they descend into deep basins, at another they ascend hills several hundred feet high.   Their rate of motion is very slow, and slower in winter than in summer, being sometimes only a few inches, and sometimes two feet or more in the course of the day, but it never ceases entirely (Fig. 15).

The fresh additions of snow which it is constantly receiving at its upper end are for ever pushing it on, urging it down the steep slopes and more slowly up the hills, and the motion is helped by the expansion and contraction of the ice with each variation of temperature, day and night, summer and winter.   Every time it expands it must creep onward, be it ever so little, and when it contracts again it

---

* Powdered glass looks white for the same reason.

Fig. 15.—THE MER DE GLACE, SWITZERLAND.

cannot retreat up the slope against the enormous weight always pressing it downwards and onwards.

Then, too, it seems probable that the freezing and consequent expansion of the molecules of water, which must drain into it whenever the surface is ever so slightly melted, also help to urge the glacier onward.

The glacier's motion, like that of a river, is greatest on the surface, and greater in the middle than at the sides ; and what with the strain resulting from this unequal motion, and the extremely rough uneven character of its bed, its surface is also rough, and rent with cracks and fissures of all sizes, from a few inches to several feet across.

Looking down upon it from a height, we should generally see on either side the glacier a dark line, which, on closer examination, would prove to be a mound of fragments, large and small, and huge blocks, many tons in weight, which have fallen from the cliffs and mountains bounding it on either side, and are thus being gradually carried down into the valley.   Thousands of tons of rock and rubbish are continually falling from the heights above ; and when two glaciers meet, as not unfrequently happens, the two nearest "moraines," as these rubbish mounds are called, join together and form a central moraine, often twenty or thirty feet high.   The three moraines then travel on together to the end of the glacier, where the ice melts and drops them, forming a "terminal moraine," perhaps eighty or 100 feet high.

But glaciers, like rivers, are "dust-makers," as well as "dust-carriers," for the joints which exist in all rocks, in a greater or less degree, make it easy both for running water and ice to force the blocks out of their places ; and then, besides the immense heaps of rubbish which the glacier carries on its surface, large quantities also fall into its cracks and fissures, and, being jammed in between the ice and its bed, are pressed against the rocks by all the weight of the mass above.   These fragments of stone are, in fact, the glacier's tools, which it holds fast with more than a giant's grip and strength, and with which it either smooths and rounds the rocks over which it passes, or else scores them with deep grooves and ruts.

In the summer, when the glacier shrinks away from the sides of its bed, it is possible to creep in below the ice and to see both the long scratches made on the rocks and how finely these have been smoothed and polished by the sand, mud, and smaller stones, which result from the perpetual grinding of this mighty millstone.

Some of the great Arctic glaciers, which are often two or three thousand feet thick, must exert enormous pressure on the blocks and sharp-edged fragments of stone imprisoned beneath them, and their beds are accordingly in some places as smooth as a polished agate, and in others are covered with grooves, which in time become deep furrows.

Streams of water flow beneath every glacier, and gushing forth at its foot, densely charged with the finest mud, form the source of many a river; but the pebbles conveyed by a glacier stream differ from those of other streams, in that they are usually angular; and the glacier is not nearly such a neat workman as the river, for instead of sorting its load, dropping the large pebbles here, the smaller there, the gravel in one place and the fine sand in another, the glacier just drops its immense piles of sand, grit, stones, huge slabs and rocks all together, and heaps them up anyhow into one great mound.

In other respects the action of the glacier is so like that of the river that, but for the peculiar tokens of its presence in the shape of rounded, scratched, and polished rocks, there would often be some difficulty in deciding which of the two had been at work.

Some of the creeks of South Australia, for instance, which have perpendicular walls of tremendous height, bear so strong a resemblance to the fjords of Norway and South America, that, but for the absence of these tokens, one might suppose them to have been excavated by the same workman. The Australian fjords, however, are really cañons, and are due to the action of rivers and torrents, whereas the true fjord has been carved by ice.

The perpetual grinding of the glacier mill-stone against the rocks which produces the "stone-meal," as it is called, naturally deepens its bed year by year; and in the course of centuries, if the supply of snow continues, it will scoop out deep channels with perpendicular cliffs. Then, if such a change of climate should take place as has occurred many times in the earth's history, the glacier will either melt away altogether or shrink higher up among the mountains, and its former bed will become a valley with, perhaps, a glacier stream running at the bottom, and here and there some of the rounded rocks, which, from their fancied resemblance to sheep, the Swiss call "*roches moutonnées*" (Fig. 16).

But if the glacier terminated on the coast instead of inland, something else might happen; for if the land sank— a thing which has often taken place—then, as the glacier retreated, the sea would flow in and occupy its bed, and instead of a valley there would be a fjord. The wonderful series of fjords by which the coast of Norway is broken, and those of Patagonia, British Columbia, Greenland, Kerguelen Land, &c., have all been thus slowly engraved by glaciers.

During what is known as the Glacial Period, a time when the climate of the Arctic regions seems to have prevailed over the greater part of the earth, enormous glaciers, whose moraines still exist, stretched from Patagonia across Brazil to Pernambuco, and as far as the equator. Traces of large

Fig. 16.—SHEEP ROCKS.

glaciers have been found also in Nicaragua, and, indeed, great part of both hemispheres was covered not merely with glaciers, but with an ice-cap, such as is now confined to Polar latitudes. During that wintry time England, too, had her snow mountains, and was for the most part covered with ice. Certainly her valleys were occupied by glaciers, one of which extended from Lincolnshire to within a few miles of London, and, gathering up specimens of the various rocks over and between which it passed, dropped them in a heap at Muswell Hill and Finchley.

Glacial drift of this kind has been found all over the northern part of Europe and America, and one of the huge old glaciers which then descended from Mont Blanc, filling the whole valley of Aosta, a hundred miles in length, has left behind it a "dust-heap," or moraine, which is 1,600 feet high, and measures sixty miles in circumference.

What was the effect upon the ocean of the withdrawal and locking up of this vast quantity of water can only be guessed; but its depth must certainly have been reduced considerably, some say by 600 feet, and some by as many as 1,000.

And when the great thaw came, and the water was set free, there is good reason to believe that at least one continent was not merely inundated, but altogether swallowed up, and that what we call the West Indian Islands are really just the highlands and mountain tops of an unknown region which then disappeared beneath the waves, and may have given rise to the various fables about the beautiful enchanted land beneath the sea, which is variously known as Atlantis, Tir-na-n-oge, &c.

We have still to say something about icebergs as dust-carriers.

True icebergs, according to Professor Nordenskjöld, are those which rise more than 300 feet out of the water, and are found only where the bed of the glacier is so steep and uneven, and its motion so rapid, that it is really split up into bergs long before it reaches the sea (Fig. 17). Even the great Humboldt glacier of Greenland, which is sixty miles broad and ends in a cliff of ice 300 feet high, could not send out such icebergs as these, for every foot of ice above the water must be balanced by from seven to nine feet below it, so that the whole height of an iceberg showing

Fig. 17.—DIAGRAM OF THE " INLANDS ICE," GREENLAND, EXTENDING INTO THE SEA, AND ENDING IN A STEEP FALL FROM 100 TO 200 FEET HIGH, FROM WHICH ICEBERGS ARE BREAKING OFF. (*After Nordenskjöld.*)

300 feet, might be between 2,000 and 3,000 feet. There are, however, glaciers in the Far North which are quite capable of sending out mountains of ice of this size; and as, owing to the extreme severity of the climate, which no rock can withstand, the Arctic glaciers are usually very heavily laden, those icebergs which break from the corners, must often carry away large loads of rubbish.

Yet it is said that, though sometimes loaded with beds of earth and rock, weighing, as has been conjectured, from 50,000 to 100,000 tons, generally speaking, Arctic bergs carry no load ; and of the numerous icebergs encountered by the *Challenger* in the Southern Seas—as many as forty being on one occasion, visible at once—not one in a thousand seemed to be carrying even mud. It must be remembered, however, that but a small part of an iceberg is visible, even when it consists of nothing but ice, and the more heavily it is loaded the deeper it must float ; besides which, large quantities of fragments, and even great rocks, might be concealed in the upper part of the bergs by the heavy falls of snow which they frequently receive after setting out on their voyages.

It is certain, however, that icebergs do act as "dust-carriers," for besides the fact that gigantic boulders are at times seen embedded in them, fragments of rock have been dredged up from the bed of the ocean, which could have been brought there only by floating ice.

Besides the monsters to which some people would restrict the term iceberg, there are other floating masses of ice which vary from a few yards to a mile in circumference, and sometimes far exceed these dimensions (Fig. 18).

On the coast of Tierra del Fuego, where almost every arm of the sea for 630 miles terminates in "tremendous and astonishing glaciers," the crash which they make as they break off into the sea, is like the "broadside of a man o' war," and in Eyre's Sound Mr. Darwin saw as many as fifty bergs floating away at once, one having a total height of at least 168 feet, and some being loaded with blocks of considerable size. In the southern seas, masses of ice from a mile to seven or ten in length are met with, and in 1854 an enormous group of icebergs was seen, locked together and forming a great hook sixty miles long and forty broad, no part of which, however, rose more than 300 feet above the sea.

Fig. 18.—BROKEN-UP BERGS.

Besides the icebergs born of glaciers, there are others of a different origin, for the sea itself often freezes in the Arctic regions, along the base of the lofty cliffs, where the water is less salt, owing to the large quantities of snow which drift into it from the shore. At low water the ice thus formed often freezes to the bottom, whence it is called " anchor " ice, and remains there glued even when the tide rises, growing thicker and thicker. When at last it floats away, it is sufficiently massive to carry with it, not only large quantities of boulders and stones, but also the anchors, cables, &c., of the fishermen, which have chanced to be embedded in it.

Sometimes, where the water is deeper, these flat masses of ice, though prevented by the rise and fall of the tide from adhering to the shore, yet remain near enough to receive drifts of snow and the waste from the cliffs, and frequently grow into islands many leagues in length and of great thickness.

Icebergs often travel long distances before they melt away. Those from Baffin's Bay come as far south as the Azores, and those from the south come within a short distance of the Cape of Good Hope, so that a large area of the sea-bottom must be strewn with the loads they drop annually, and should it ever rise to the surface, will be found covered with " drift," consisting of gravel, stones, and boulders, which have no connection with the rocks on which they rest, and are scattered about helter-skelter, with no more attempt at sorting or stratification than if they had been " shot " from so many dust-carts.

To a small extent, icebergs are dust-makers as well as carriers, for when they strand, they groove and polish the rocks in true glacier style, and, off North America, they push pebbles and sand before them, leaving the submarine rocks quite bare.

CHAPTER VI.

DUST-MAKERS—EARTHQUAKES AND VOLCANOES.

How Sea and Land have changed Places—The History of the Petrified Firs—Perpetual Motion—Subterranean Heat—Earthquakes and Earthquake-waves—Wear and Tear—Volcanoes, their Ash and Dust Heaps—Double Work done by Earthquakes and Volcanoes—Hot Springs—Mammoth Springs.

WE referred in the last chapter to the possible rise of the bed of the Atlantic (p. 76), at some future period, above the waves, and we have now to see how this might be brought about.

It is a well-known fact that sea and land have many times changed places, and that by far the greater part of the rocks composing the earth's crust must, from their character, have been formed under water.

It has also been ascertained by careful observation that, at the present time, Norway and Sweden are quietly rising higher and higher out of the German Ocean, at the rate of three feet in a century ; and on the eastern coast of South America there are large beds of shells which have been raised, some a few feet only, others as many as three or four hundred feet above the sea, in what geologists would call "quite modern times;" while at Santa Cruz a rise of at least 1,400 feet has taken place since the time when the great boulders with which the plains are dotted were dropped by glaciers or icebergs.

On the western coast of America the evidence as to the way in which sea and land have changed places is even more striking, and shows that the vast mountain chain, variously called the Cordillera, Andes, and Rocky Mountains, which apparently stretches in an unbroken line from Tierra del Fuego to the Arctic Circle, with pinnacles here and there reaching a height of 20,000 feet, has risen and sunk again certainly once, and almost certainly twice, in the course of its history.

In the Uspallata range, which is separated by a narrow plain from the main mass of the Cordillera, there stands a group of snowy-white columns, a few feet high, which have a weird, ghostly look about them, and are evidently the trunks of trees which have been petrified and converted, some into flint, and others into coarsely-crystallised spar. Now, these stone fir-trees have had a wonderful history. They must, of course, have grown upon dry land, but below them are several thousand feet of rock, which could have been formed only under water; therefore the land hereabouts must gradually have sunk lower and lower, until it was many thousand feet below the surface of the sea. Then, when these new beds had been formed, it rose again, and when the soil had been prepared for them, a beautiful group of trees sprang up and flourished on the shore of the Atlantic, which then washed the foot of the mountains, though it is now 700 miles away. The trees grew to maturity, and then the land began to sink again; but this time it was let down to an enormous depth, much greater than before, for the firs were buried under beds of sediment as thick as those upon which they stood, and besides this, streams of lava from a submarine volcano flowed over them, one being a great mass of black basalt, 1,000 feet thick. After being thus buried, the trees were once more raised, and this time hoisted up 7,000 feet above the sea, and now, by the wear and tear of time, they have been exposed to view and stand high up in the mountains and hundreds of miles inland, like ghosts of their former selves.

Similar risings and sinkings of the land have gone on, and are still going on, more or less, all over the world, so that it is quite true, as has been said, that "nothing, not even the wind, is so unstable as the crust of this earth."

As to the causes of this perpetual motion scientific men are at present by no means agreed. On whatever part of the earth's surface we may be standing we are something less than 4,000 miles from its centre; but the deepest mine in this country has not quite been carried to the depth of even half a mile, which is a mere scratch in comparison. What would any one know of a cocoanut if he had but scratched its shell? And yet a scratch on a cocoanut shell would be proportionately far deeper than the deepest cutting we have yet made in the crust of the earth.

We may argue, indeed, that because the mercury in a thermometer is found to rise on an average one degree for each sixty feet that it is carried below the earth's surface, therefore, at the depth of but a few miles the heat must be so intense as to melt any rock with which we are acquainted; and if we can also prove that the heat *goes on* increasing at the same rate throughout, our argument may be reasonable as far as it goes (though even then there are other matters to be considered), and we may be further justified in adding that it is the changes taking place in this molten mass which cause the earth's crust to rise and sink.

But we cannot prove anything of the kind, and our knowledge of the two thousand feet or so which we have partly explored at a few scattered spots does not give us any information as to what goes on, or as to the temperature which may prevail at the depth of even a hundred miles.

All that we can say is that, whatever may be the state of the interior of the globe, the heat at some spots, and these comparatively near the surface, is great enough to melt rock, and to keep water not only hot, but in many cases boiling.

It is possible, therefore, that the slow movements in the earth's crust may be due to the cooling and re-melting and consequent contraction and expansion of miles of rock; but whether this is really so we are quite unable to say.

Neither are we much better off when we try to find out the causes of those sudden and violent movements to which we give the name of *earthquake.* These are so explosive in their character that one might at times imagine them to be occasioned by the sudden conversion into steam of water which has filtered downwards until it has come in contact with heated rock; but whatever the cause "an earthquake," says Prof. Huxley, "is just such a disturbance of the ground as would result from a sudden shock or blow given upwards in the interior of the earth,"* by which tremors may be communicated in all directions through the solid rock.

The more solid the rocks the more they would feel the jar, and the better able they would be to pass it on to the rocks above, whereas in a bed of loose gravel it would be almost extinguished.

If a number of ivory balls are placed in a row touching one another, a tap given to the first will be felt by all, and the last, having nothing to keep it in its place, will fly off. In a similar manner a shock of earthquake has been known to be so violent as to send paving-stones flying into the air, with such force that they turned a complete somersault.

It is often found that an earthquake has caused the area affected by it either to sink or rise, more generally the latter, to the extent of several feet. The memorable earthquake of 1835, which shook the western coast of South America, raised the land round the Bay of Concepcion two or three feet at one blow, and upheaved a rocky flat off the island of Santa Maria, which was left with its beds of gaping mussels hopelessly stranded, ten feet above high-water mark.

More havoc, too, was wrought in the island of Quiriquina by this earthquake than would have been accomplished by the ordinary wear and tear of a century. Its effects were felt far and wide, and had it occurred in Europe the whole continent, from the North Sea to the Mediterranean, would, Mr. Darwin says, have felt the tremendous

---

* The greatest depth at which the shock originates appears, according to Mr. R. Mallet, to be forty miles, and the smallest eight.

G

jar, and a large tract on the eastern coast of England would have been permanently upheaved.

There is probably no place upon the face of the globe which is entirely free from the vibration of earthquake shocks, and these, therefore, contribute much, both directly and indirectly, to the wear and tear of the earth's crust ; for besides the destruction caused in severe earthquakes by the trembling and rending of the rocks, we must not leave out of sight the often worse destruction caused by " earthquake waves," when the disturbance takes place near the coast. The sea then feels the shock as well as the land, its bed is frequently raised several feet, as by a sudden jerk, which, of course, powerfully affects the water above, and enormous waves are produced, which rush upon the shore, carrying everything before them. In 1835 the great wave which swept along the South American coast, left not a house standing in Concepcion or Talcahuano, almost washed away even the ruins of the latter place, and, breaking at the head of the bay in a fearful line of white breakers, rushed up to a height of twenty-three feet above the highest spring-tides. Its force was so great that it moved a gun and gun-carriage, weighing four tons, a distance of fifteen feet, but it advanced at such a deliberate pace that the people had time to run up the hills out of its way.

After an earthquake which occurred in Japan in 1854, the waves continued to come and go from 10 a.m. to 2.30 p.m. Not a house was left standing in the harbour of Simoda, many junks were carried inland, one of them as much as two miles, and a few hours after the disturbance in Japan several well-marked waves had reached the coast of California.

The destruction occasioned by earthquakes is obvious enough, but even the gradual sinking and rising of the land contributes much to the wear and tear of the coast, since by this means different parts and fresh surfaces are brought within reach of the waves. The straits between islands, for instance, are worn deeper and deeper as the land gradually rises, until, when it has been lifted quite beyond their reach, the straits become mountain passes, connecting one valley with another.

It has been said that whatever the condition of the interior of the earth, we have positive evidence that at some spots not very far beneath the surface, the heat is great enough to melt rock, for many a subterranean disturbance, which begins only with the quaking of the earth, ends with the pouring forth of liquid matter.

A crack is made in some weak place, through which large volumes of steam and other vapours are forced up, with showers of red-hot ashes and streams of molten rock.

The fall of these materials round the mouth of the hole forms a cone-shaped mound, called a volcano, which usually has a funnel-shaped opening, or crater. The pipe, or chimney, which leads down into the interior of the earth, has a hard stone lining, formed by the melted rock or lava, which cements the loose ashes and cinders into a compact mass wherever it comes in contact with them.

It seems probable that much of the force by which volcanic matter is driven to the surface, and shot up into the air, is due to the conversion into steam of water which has found its way down to the molten rock below; and whether or no this be the sole source of volcanic energy, it is certain that steam is poured forth in large quantities at the beginning of an eruption, and that with such violence that any fragments of rock which may have accumulated in the throat of the volcano are hurled into the air with much force. Masses of rock, some nine feet in diameter, have been hurled fifteen miles by the great volcano of Cotopaxi, and even larger blocks than these have at times been sent flying several miles, or shot up to a height of 6,000 feet.

Ashes—that is, fragments of lava or partly-melted rock, which have been so splashed about as to fall in spongy-looking drops—are poured forth by volcanoes in vast quantities, and are frequently broken up into particles so fine as to be nothing more than dust, which fills the air, and plunges the whole neighbourhood for miles round into darkness.*

The most remarkable eruption which has occurred in our times is that of Krakatoa, which stands upon a fissure running across the Straits of Sunda, and until 1883 had been quite quiet for two hundred years. Many earthquakes, however, had recently taken place, and it may be that a larger quantity of water than usual was consequently admitted into the depths below. The eruption began in May, 1883, when the sea for ten or twenty miles was covered with drifting pumice, through which a ship cut her way with as much noise as if it had been thin ice; the volcano continued more or less active for the next three months, and the worst out-

* See Belt's "Naturalist in Nicaragua," for eruption in Coseguina.

break of all occurred on the 26th and 27th of August, when an incessant rumbling was heard accompanied by short, loud reports, as if from heavy guns. No shocks of earthquake were observed, but the blast of air produced by these explosions was so violent that walls were rent at a distance of some 500 miles.

The island of Krakatoa consisted of three peaks, of which the most lofty was 2,700 feet high; but the whole of the northern part with two craters has disappeared in the sea, and half the remaining peak has sunk likewise, having been cut in two from the very summit, so that it now forms a cliff between two and three thousand feet high. A great wave, caused no doubt, by the sudden subsidence of this peak, started from Krakatoa with a height of twenty-seven feet, dashed upon Java and Sumatra, and opposite Anjer, in the narrow throat of the Straits, rose to from forty to a hundred feet, sweeping the shore of thousands of its inhabitants. The effects of this wave were felt on both coasts of America; ashes thrown up by the volcano fell over an area almost as large as Norway and Sweden together; dust fell to the depth of two inches upon a vessel 1,000 miles off, and another vessel, which was near the Straits, passed masses of floating pumice seven feet thick, which in some places were so extensive as to impede navigation.

And here we must remark that, although we have hitherto considered them almost exclusively in their destructive character as dust-makers, yet earthquakes and volcanoes, like many of Nature's other labourers, do double work, and build up as well as pull down. Indeed, it is no exaggeration to say that without them the whole of the dry land on the globe might be buried beneath the ocean; for as the rivers are perpetually engaged in wearing it down and carrying it away, they would in time reduce the level of the land everywhere to that of the sea, were it not for the compensating earthquake force, which whether working suddenly or gradually, acts on the whole in the opposite direction.

Volcanoes also have their share in repairing the waste of the earth's surface, and though the island of Krakatoa was reduced by the great subsidence to less than a third of its original area (twenty square miles) it has since been increased more than three square miles by the addition of volcanic matter.

The amount ejected by Krakatoa was, however, small compared with that thrown up or poured forth by many other volcanoes.

Most of the islands of the Pacific are of altogether volcanic origin, hills and islands of volcanic matter have frequently been raised to a height of several hundred feet in the course of a few hours, and lava has been poured forth in such voluminous streams as to form beds of vast extent and many hundreds of feet in thickness.

Whether the molten rock which we call lava has itself been formed from something else and has previously existed in a solid state, we have no means of knowing; but there are evidently vast stores of it, for the space from beneath which volcanic matter has been ejected by volcanoes in the Cordillera alone measures 720 miles one way and 400 the other, showing, as Mr. Darwin says, the existence of a subterranean lake of lava nearly double the size of the Black Sea.

A stream of lava soon cools and hardens on the surface, but the great mass within retains its heat for years. In 1759, the Mexican volcano Jorullo poured forth a perfect sea of lava which completely filled the beds of the two rivers Cuitemba and San Pedro for some distance. Both streams disappeared on one side of the vast expanse of molten rock, but made their way beneath it and reappeared on the opposite side as permanent springs pouring forth large bodies of very hot water, which retained their heat for many years after the eruption. At the present day the water is but a few degrees warmer than the air; but for a hundred years or so large quantities of steam continued to issue from the volcano.

All the Azores are islands of volcanic origin, and in one of them, San Miguel, there is an immense crater, which in former times was no doubt a lake of boiling lava, but is now a wonderfully green and fertile hollow, called, from its shape, the Cauldron of the Seven Cities, and containing two lakes and many villages, whose white houses, meadows gardens, and corn-fields, lie at a depth of 1,500 feet below the lip of the crater.

San Miguel is still shaken by earthquakes from time to time, and although there are no more eruptions, the underground fires still burn well enough to keep the water in a number of springs always boiling. These springs all empty themselves into one small stream, which retains its heat for several miles and carries down into the sea a great variety of dissolved minerals. Hot water and steam are more powerful solvents than cold water, and being usually charged with large quantities of carbonic acid, act very powerfully upon limestone, potash, soda, iron, magnesia, manganese, and various compounds of silica; and then, when the water is impregnated with these, especially with the salts of potash, soda, lime, and magnesia, its dissolving power is again greatly increased, and far surpasses that of water which is merely charged with carbonic acid.

It is no wonder, therefore, that hot springs contain a large proportion of mineral matter, and contribute much to the wear and tear of the rocks through which they pass.

At Terceira the hard lava rock from which the hot springs issue first becomes earthy and covered with red speckles from the rusting of the iron which it contains; then it gradually turns soft, and at length even the glassy crystals of felspar dissolve, and the rock is converted into clay, some of which is quite white from the removal of the iron, and looks like the finest prepared chalk, while some is bright red from the accumulation of iron oxide or rust.

In England we are far removed from any active volcano, yet we have at Bath hot springs, whose heat, though not so great as those of Aix-la-Chapelle and other places, is still rather startling when we come to consider what it means. The Bath water is not remarkable for the amount of mineral matter which it contains, yet Professor Ramsay has calculated that if solidified, it would in one year form a pillar 140 feet high and nine feet in diameter. All this is now carried away unseen to the Avon, and by the Avon to the sea.

The most wonderful hot springs yet known are those called the Mammoth springs of the National Park of Yellowstone, a tract of country in the heart of the Rocky Mountains. At the northern end of this "park" which has an area of 3,575 square miles, rises a greyish-white cliff about three miles in length, which on a nearer approach resolves itself into a series of terraces, the steps of which are occupied by numerous natural basins, some gigantic, some minute, and all overhung by clouds of silver vapour. Some of the terraces are several feet wide, others quite narrow, and the steps also vary in height from ten feet to an inch or two.

The top level is 150 feet wide, and here rises the largest spring in a basin forty feet long and twenty-five wide. The principal springs occur on the first ten terraces, and the clear, blue water, flows down from various openings in the rims of the basins, whose general shape is oval, the edges being scalloped in graceful curves with wavy frill-like borders, frequently adorned with pearly knobs from the size of a pin's head to that of a hazel-nut, or with wonderful incrustations resembling coral, moss, feathers, butterflies' wings, &c. The prevailing colour is a rich cream, but there are touches of bright sulphur-yellow, delicate pink and salmon, vivid scarlet, green, rose, crimson, purple, and brown. This beauty is produced by very simple means, and the marble basins, ornaments, and all, are merely carbonate of lime coloured by iron and sulphur, and have been deposited by the water.*

The water of the Mammoth springs, which of course still contains a large quantity of mineral matter, notwithstanding these deposits, flows by several channels into Gardiner's River, and so to the ocean (Fig. 19).

---

* There are similar terraces of basins, also formed by hot springs, at Te Tarata, New Zealand ; but these, though as white as marble, are deposited by water containing silica instead of carbonate of lime.

Fig. 19.—Mammoth Hot Springs on Gardiner's River, Wyoming.

CHAPTER VII.

DUST-MAKERS—VEGETABLES AND ANIMALS.

Lichens the first Soil-makers—How they eat into the Rocks—Vegetable Acids—Roots, their Length and Number—Minerals required by Plants—Flint in Grasses, Dutch Rushes—Mechanical Power—Burrowing Animals, Worms, Ants, Marmots, Birds, Boring Mollusks.

WE have seen from the last chapter that a stream of lava continues intensely hot for a long time, and may emit vapour, and even go on creeping slowly down the mountain side for years after it has been poured forth.

Long after it has ceased to move, and centuries after it has become perfectly cold and solid, the surface of the stream will remain fresh, smooth, and glossy, effectually resisting all attempts both of air and weather to convert it into dust, as we may see by the lava streams of Ascension Island.

Even lava has to yield at last, however, and that to workmen whom we should at first sight be inclined to call very insignificant, nothing more, indeed, to all appearance, than mere *stains*, such as one sees upon a brick wall. These discolour the surface here and there, and upon examination with the microscope prove to be vegetables, of a very humble kind it is true, but still vegetables, Nature's gardeners and first soil-makers, which will prepare the way for more important plants. Lichens, as they are called, are to be found everywhere; no climate, hot, cold, damp, or dry, and no soil seem to come amiss to them, for they will grow loose on the surface of the sand in Peru, and attach themselves to the dry bones of mules which have died by the wayside. They are invariably the first plants to make their appearance, whether upon lava or upon the rocks of islands newly raised above the sea, and no rock is too hard for them. The multitudinous spores by which they are propagated are for ever floating in the air, and being furnished with a gummy fluid are able to attach themselves to the barest, smoothest surface ; and when once they have gained a footing, they are simply irresistible.

They flourish on granite, slate, lava, and in Berkshire (Mass.) even the white quartz hills are covered on their moister slopes with large patches of a leathery lichen, which adheres so firmly that it can hardly be detached from the stone.

The first lichens to appear are, as has been said, mere stains ; but, by the growth and decay of successive generations of these, a thin film of soil is formed, upon which larger kinds take root in their order, and at last one may see rocks, or old tombstones, covered with a crust of lichen an inch or more thick. On this crust mosses begin to grow, and they help on the process of decay by keeping the surface moist and sending their roots down into the stone ; then insects collect and feed, die and decay, and thus the mineral matter of the rock is not only reduced to powder but mixed with organic remains, without which it would be quite unable to support the higher orders of plants.

The mosses, as they grow thicker and thicker, keep the air from the rock, and thus to a certain extent protect it ; on the other hand, they also keep the surface moist, and as moist surfaces absorb more carbonic acid than dry ones, this also helps forward the decay to some extent, even though the other enemy—frost—be kept out.

Now, how do the lichens manage to wear away the rocks, since, though low down in the scale, they are certainly plants, and plants can live only upon liquid food ; yet these eat their way not only into bricks which have been baked, but into rocks which have been vitrified, converted, that is, into a sort of glass by the heat of the subterranean fires?

On removing a lichen from the rock, we see that it has made an impression more or less deep, and that the stone has lost so much of its substance ; and further, on burning the lichen, we find that it contains from ten to twenty per cent. of solid ash, which will not burn away, being, in fact, mineral matter abstracted from the rock.

How then has the lichen managed to feed upon stone ?

All plants have the power of forming within themselves acids of one sort or another,* and as lichens do this to a larger extent than most others, they are able to dissolve the very hardest rocks. Many of them, indeed, so abound in oxalic acid that oxalate of lime makes up half their weight, and many of the old Greek marbles are thickly encrusted with this substance from the growth and decay of lichens on their surface. Oxalate of lime is simply a combination of oxalic acid and lime ; the acid having dissolved the lime from the rock or marble, and so brought it into a condition in which the lichen could absorb it.

But though lichens are the first soil-makers and take up

* See Chap. XI.--Vegetable Scavengers.

more mineral matter in proportion to their size than any others, still, all plants contain some acid, and all, on being burnt, yield a certain quantity of ash, some more, some less, which they have absorbed from the soil and could not have done except in the form of liquid.

We have already seen something of the solvent powers of carbonic acid, and this gas plants are constantly giving off through their roots, which are much more numerous than people generally have any idea of. Rye, beans, and peas, for instance, will send down a thick mat of white fibres to a depth of four feet ; winter wheat has been known to send out roots seven feet long in forty-seven days, while clover a year old has roots three feet and a half long.

But this is not all. Besides these obvious roots, there are very many others so fine and hair-like that they escape notice altogether. Yet the plant takes up its nourishment through all and especially through these young, almost invisible roots (Fig. 20).

By way of experiment, beans, maize, and wheat, have been planted in fine quartz sand, having at the bottom plates of marble (carbonate of lime), magnesian limestone (carbonate of lime and magnesia), gypsum (sulphate of lime), and glass.

Being kept well watered, the seeds soon began to sprout and send out roots, and when they had reached the plates below and could get no farther, they spread themselves out horizontally.

When the plates were examined after a time, the marble and magnesian limestone were found to be corroded—eaten into—by the roots, the impression of even the root-hairs being

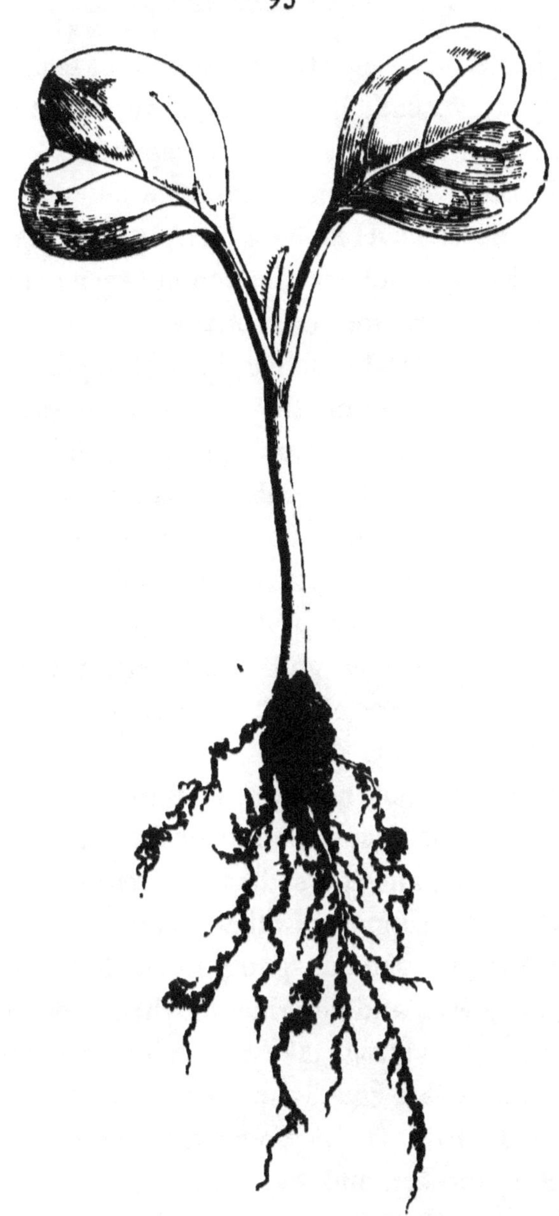

Fig. 20.—Young Turnip Plant Nine Days from Date of Sowing, showing how the Root-hairs are closely covered with firmly-adhering Particles of Earth (2½ *times natural size*).

distinctly visible. No effect seemed to have been produced on the glass or gypsum ; but that, after all, might only mean that it was as yet too slight to be perceptible.

Smooth pieces of limestone are often found in meadows with their surfaces covered by a perfect network of small furrows, which on careful examination are seen to correspond exactly with some tiny root or rootlet.

Lupins are especially active in decomposing mineral matter, and for this reason they are sometimes, as in the Azores, planted with corn and ploughed in as manure. Three lupins planted in powdered sandstone have been found to take up three-fifths of a grain of mineral matter ; an equal number planted in powdered basalt took up three-quarters of a grain. Three peas took up rather less basalt and much less sandstone, and buckwheat, vetches, wheat, and rye, considerably less of both ; but all showed a marked preference for the basalt.

Their work did not, however, end with what they had absorbed, for it was found that they had dissolved more than they had used, and the soil was, therefore, so much the better fitted for the support of other plants.

All plants require sulphur, phosphorus, flint, iron, potash, soda, lime, magnesia, and chlorine, for their proper development; and though they may take but an infinitesimal amount of some of these, all are equally necessary, and the absence of any one would prove fatal. Some take more of one and some more of another, and even the different parts of the same plant may take the minerals in different proportions. Thus grasses and all the varieties of corn take up much more flint than turnips and cabbages do; and while the ash of horse-chestnut bark contains 12 per cent. of potash and 76 of lime, the ash of the wood yields twice as much potash, that of the leaf-stem nearly four times as much, and that of the flower 61 per cent. of potash and 13 of lime.

All the cresses, on the other hand, contain a large proportion of sulphur, and though a crop of mustard may be grown upon damp flannel, it is not to be supposed that it lives and flourishes on air and water only; for the water, unless it be distilled, is quite certain to contain some mineral matter, and from this source the plants derive sufficient food for a time. If grown in pure water, they will sprout and even grow at first; but as soon as they have used up the matter contained in the seeds, they will die; and if they be burnt, the remaining ash, when weighed, will be found to be only the same in amount as that contained in an equal number of seeds, showing that they have not been able to find anything in the water.

Every crop, therefore, takes a certain amount of mineral matter from the soil,* and as one takes more of one thing and another more of another, the farmer varies his crops that the soil may not be exhausted. Sometimes he will let it lie fallow altogether, that the carbonic acid and ammonia washed down by the rain may dissolve the minerals and prepare them for plant-food, and the same end may be more quickly attained by the growing and ploughing in of green crops. But, of course, neither plan will answer unless

* In general, probably 99 per cent. of the soil without water does not contribute to the support of vegetation. The hay crop takes up the largest amount of mineral matter, but reckoning it at 2½ tons to the acre, it takes only 400lbs. The weight of the soil, one foot deep, is 4,000,000lbs. per acre. (S. W. Johnson.)

H

the minerals are already in the soil ; and if they are absent the farmer himself must supply them in the shape of bones, ashes, &c.

Many grasses, especially the stiff harsh kinds, and all varieties of corn, contain much silica or flint. What they want it for is not clear, for it does not seem needed to give strength to their stems, as has been supposed ; but they cannot do without it. In barley straw more than half the ash is flint, in winter wheat 41 per cent., in meadow hay 29 per cent.*

Among trees, while the willow and oak contain very little flint, but a great deal of lime, the Scotch fir takes as much lime as the willow, and twenty-one times as much flint ; and the Cauto tree of South America has bark as hard as soft sandstone from the quantity of flint it contains, and the natives of Trinidad use its ashes in place of sand, with the clay of which they make their pottery. The smooth glossy rind of the bamboo, which contains 70 per cent. of flint, will strike fire with steel, and the same substance collects in hard lumps like opal in the joints of the stem.

There is flint again in the hairs of the nettle, in hemp, and in hops, and all the best vegetable weaving fabrics, except cotton, contain a great deal. Of all our native plants, however, none contain more than the horse-tails, formerly

---

* The proportion of ash of *all* sorts, though not invariable, is about as follows :—

	Per cent.		Per cent.
Field beet	18·2	Pea straw	7·9
Red clover	6·7	Pea grain	2·7
Wheat straw	5·4	Fir bark	2·0
Wheat grain	2·0	Fir wood	0·3

used for polishing by cabinet-makers and metal-workers, and still made use of in some parts of the country for scouring pots and pans. Half of their ashes, and in some varieties more than half, consists of flint, so that they are a sort of natural sand-paper, and were at one time largely imported under the name of " Dutch rushes." If examined through a microscope, their cells, regularly arranged lengthwise in ridges, may be seen encrusted with silica ; and if the whole plant be placed in nitric acid, all the soft parts will be eaten away, while the flinty skeleton will remain entire.

Living structures, whether animal or vegetable, possess extraordinary powers of acting on mineral matter, and a humble lichen will produce more effect on a piece of rock with the sulphuric, oxalic, or nitric acid it contains than an equal amount of acid would do if otherwise applied ; nor would a dead lichen, however acid, be able to do what a living one does.

What is true in this respect of lichens, is true in their degree of all plants ; and when we consider to what a depth many roots must penetrate, and that wherever they go they effect some chemical change, we can readily understand that, as dust-makers, they are too important to be passed over.

Moreover, the chemical effects of vegetation are not all that is to be thought of. The mechanical effects must also be taken into account.

Few plants are more fragile and delicate-looking than a maidenhair fern, yet one knows that its roots are strong enough to crack the pot if they have not room enough ; weighty kerb-stones have been seen to be forced completely out of place by the tender blades of grass growing between them; and Canon Kingsley mentions having seen a large flat stone raised up in a single night by the growth of a crop of tiny mushrooms.

Then, again, we read of a nut tree springing up in the centre hole of a disused mill-stone, whose stem grew and increased in size until it entirely filled the hole; whereupon it gradually raised the stone from the ground, until the huge mass, some five or six feet in diameter, was lifted up eight inches all round, and was supported in the air by the stem only. The tree attained the height of twenty-five feet, bore excellent fruit, and was killed at last, not by the weight of the stone, but by the tightness of its embrace, which stopped the flow of sap. This is certainly "a striking example of the tremendous powers of Nature;" and since trees and plants grow as vigorously downwards as upwards, their roots must certainly help to break up the rock, not only by causing decomposition, but by forcing their way into joints and cracks and then gradually widening them. The island of Aldabra, north-west of Madagascar, for instance, is being reduced and destroyed by the silent inroads of the mangroves, which grow along the base of the cliffs and have eaten their way into the rock in so many directions that the island is completely riddled by creeks of their making.

But we must now pass on to more intelligent burrowers, foremost among whom comes the great army of worms.

It has been calculated that there are as many as 53,767 worms in each acre of garden ground, and about half that number in cornfields. Nine burrows, and sometimes many more, are usually found in two square feet of garden soil;

and it is computed that, all over the 32,000,000 acres of cultivated land in which worms are able to live in Great Britain, ten tons of earth are brought to the surface in each acre by the worms alone every year. Moreover, various acids, called humic, are generated by the digestion as well as decay of the vegetable matter which forms so large a part of their food ; and as these seem to be even more powerful than carbonic acid, they must play an important part in the disintegration of the rocks. Worm burrows being frequently five or six feet long and even longer, some small amount of acid at least must be carried down to these depths, there to act upon the underlying rock or fragments of rock ; and as all the mould is in constant though slow movement, fresh surfaces must be continually exposed to the action both of the humic acids and the carbonic acid of the soil.

The green streaks, sometimes seen in red marls* or sandstone, are caused by the decay of vegetable matter ; humic acids have been formed, and these have robbed the red oxide of iron, to which the marl owes its colour, of some of its oxygen, leaving it green or bluish-green ; and, as worms drag enormous quantities of leaves into the ground as linings for their burrows as well as for food, they must greatly promote the formation of these powerful acids.

Worms also frequently undermine and even penetrate the walls of old buildings, and by thus causing them to sink have by degrees helped to bury them. No building is safe from their burrowing, it is said, unless its foundations be carried down to a depth of six or seven feet.

---

* Marls are a mixture of lime and clay with little or no sand.

Worms are found in all moderately damp countries, and wherever they are they help to wear away the rocks and make "dust," not only in the ways already mentioned, but also by swallowing a large amount of earth and even small stones, the latter of which are ground and reduced in size by the process of digestion, while the former is rendered so fine as to be more easily washed away by the rain.

In some countries ants, also, do an amount of excavation which may well be called enormous considering the size of the workers. At Rio de Janeiro, for instance, a species of the Saüba ant has made a tunnel under the bed of the river Parahyba, where it is as wide as the Thames at London Bridge; at the Magoary rice mills, near Parà, they once pierced the embankment of a large reservoir; and upon fumes of sulphur being blown down some of the main entrances to their colonies, the smoke was seen to issue from a great many outlets, one of which was seventy yards away.

Among the larger burrowers may be mentioned rabbits, moles, marmots, sand-martins and other birds. In some parts of Tartary, the rocks are perfectly riddled by the holes of marmots, and in South America there is a small bird, called by the Spaniards Casarita, or Little Householder, which makes a narrow cylindrical hole nearly six feet under ground, at the bottom of which it builds its nest. The holes are made in any low bank of firm sandy soil, by the side of a stream or road, but occasionally the birds make the mistake of choosing a mud wall, and at Bahia Blanca Mr. Darwin saw one which they had pierced in twenty places, to the great annoyance of its owner. The wall was a low one, and the birds were constantly flying over it, but had not the wit to see that it was not thick enough to suit their purpose.

It has been already mentioned that roads paved with hard granite are quickly worn away by the constant passage of heavy traffic, but it would hardly be imagined that the passing to and fro of birds' feet could make any impression on the rocks, yet in Nightingale Island, Tristan da Cunha, these are actually smoothed and polished by the continual tread of hundreds ot penguins on their way to and from the sea.* No doubt the flinty skeletons of the microscopic plants called diatoms, which are always found in abundance in the mud about their nests, adhere to their feet and act as polishing powder.

Beneath the penguin rookeries are the holes of the prions and petrels, which are bored in all directions, the round being honeycombed to such an extent that it often gives way when human beings venture upon it ; and much the same may be said of the sandy flats about the Cape of Good Hope, only there the burrowers are moles, whose tunnels are so large as easily to admit the hand and arm.†

The boring mollusks, too, must not be entirely passed over, as, although their shells are as thin as paper, and as brittle as glass, they are able to pierce wood, limestone clay, slate, and even sandstone, to the depth of several

---

* The island is a mile square and is inhabited by about 400,000 penguins, whose rookeries cover a quarter of its area.

† In the Chilian Andes, the burrows of the little chinchilla are so numerous as considerably to increase the difficulty of travelling.

inches, and thus frequently destroy the foundations of jetties, sea-walls, &c., and ultimately cause their destruction.

Plymouth breakwater, which is constructed of hard marble-like limestone, was so much injured by their ravages

Fig. 21.—PHOLAS IN A SHELTER HOLLOWED BY IT IN A BLOCK OF GNEISS.

that it was found necessary to replace the blocks between high and low water-mark by granite, against which they are powerless.

Somewhat similar excavations are made by a species of land-snail; but this is believed by Dr. Buckland to work with its rasp-like tongue, whereas the pholas seems to have no other tool than its papery shell.   (Fig. 21.)

CHAPTER VIII.

WHAT BECOMES OF THE "DUST"—TOWNS AND CITIES.

What becomes of Nature's "Dust"—Its Amount—The Visible Part—Sand-banks and Sand-dunes, River Sands—Sandy Deserts—Sand as a Preserver—Uses of Sand—Glass-making—Sandstones and Precious Stones—Mud, Shales, and Slates—A Great Mud-heap—Effect of Pressure—Porcelain Clays and Brick Clays—A Mass of Sapphires and Rubies—Towns and Cities.

IN countries where there are no violent earthquakes or volcanoes, no tropical storms, and no glaciers, Nature's labourers make and carry away their "dust," for the most part, so quietly, and the hills look so very much the same from year to year, and even from generation to generation, that we may find some difficulty in realising that anything at all is going on. Yet we are told that, by one means and another, a mile in thickness has been worn away from the Mendip Hills; that the part of Kent and Sussex called the Wealden has been stripped of a mass some hundreds of square miles in extent, and several hundred yards in thickness; and that the district to the south of Snowdon has lost from its surface a mass of 20,000 feet, a whole mountain, in fact, as lofty as the Andes.

Not an atom of all these many million tons of rock has been lost, however, though it may have changed its appearance so much as to be often hardly recognisable, and it is when we consider what has become of it that we perhaps gain the best idea of what Nature's wear and tear really means. For "the entire mass of stratified deposits is the measure of former denudation." In England, roughly speaking, all the rocks, with the exception of the granite hills of Cornwall, Devon, and Worcestershire, are stratified, *i.e.*, have been deposited in beds or strata at the bottom of seas or lakes, or at the mouths of rivers; all have had a previous existence in some other shape, and have been worn away from some unknown land, which we may dream of as Atlantis if we will.

Rivers are the great carriers, and if the refuse conveyed by them has had a long journey, the visible part of it reaches the sea in the form chiefly of mud and sand,* which are both usually deposited within one or two hundred miles from the shore, though the finest portion may be carried farther, especially where the current is strong, and if it becomes entangled in ice, is, of course, carried much farther still.

Much of the sand remains close in shore, forming shoals and sand-banks, and much is thrown up on the beach, where it is dried by the wind, and then, unless there are cliffs or rising ground to stop it, is frequently blown inland again, and piled up into sand-hills, which drift farther and farther year by year, swallowing up houses, villages, and even forests.

This is what the sand has done on the flat coasts of Jutland, and the Bay of Biscay. The Landes, as the sand-dunes are called in France, extend from the Garonne to the

---

* The beach pebbles are, for the most part, made by the action of the waves on the coast, and though often swept along it, are seldom carried out to sea.

Pyrenees, and moving forward at the rate of sixty or seventy feet every year, have buried several villages, which were well known in the Middle Ages. In some places their progress is arrested by quite small running streams, the sand, as it drifts into the water, being carried back into the sea; but, on the other hand, they have proved more than a match for the river Adour, which they have turned nearly a mile and a quarter out of its original course. (Fig. 11, p. 38.)

In the outer Hebrides, the encroachment of the sands has been checked by planting them with the sea-reed or mat-grass, whose tough roots are often twenty feet long, and serve to bind the sand together.

In Ceylon, where the rivers flow rapidly down from lofty hills, they reach the coast heavily laden with sand and mud, which, instead of being carried any distance out to sea, are heaped in bars along the shore by the currents of the Bay of Bengal. The bars extend north and south, and at length attain such dimensions that the rivers, being unable to force their way through, are obliged to flow behind them in search of a fresh outlet.* Long embankments, from a mile to three miles broad, and forty miles long, have thus gradually accumulated, and having first been in some degree consolidated by the growth of an ipomœa or convolvulus, which sends out roots from every joint, the soil has then been fertilised by glassworts, saltworts, and other sand-loving plants, until at last it has become capable of supporting plantations of cocoanut trees.

* These barriers grow especially fast on the east of the island, to which large quantities of sand are brought by the southern current from the Coromandel coast.

Many of the river sands of Ceylon consist of fragments of rubies, sapphires, and garnets, intermixed with others of quartz and mica ; and the bed of the Manickganga in particular is composed to such a large extent of ruby sand that it reminds one of the story of Sindbad ; but, as none of these precious fragments is larger than a mustard seed, the sand is valueless except for polishing and for sawing ivory.

Among the great accumulations of sand existing at the present day must be mentioned those of the desert of Gobi, lying north of the Himalayas, concerning whose buried cities and marvellous hidden treasures many tales are told.   One of these cities, named Pirna, is said by the Chinese to have been suddenly overwhelmed in the sixth century; and an interesting account exists of the flight from another, called Katah or Kank, of a Mahometan priest, who for many successive Fridays had warned his flock ot the calamity about to fall upon the city.

In this desolate region the wells are all protected by huts, else they would soon be choked by the ever-shifting sand, which stretches away in the distance like a great sea marked by regular waves, which rise, one behind the other in rows, to the height of ten, twenty, or even one hundred feet, leaving the hard under-lying clay exposed to view between them.

The advance of the sand takes place chiefly in the spring, when the wind blows constantly from the north ; but even then it is often so gradual that people will go on occupying a tenement, whose court may be filled with sand up to the verandah, from the breaking of the sand-dune over the wall.

But though destructive from one point of view, nothing, on the other hand, is better fitted than sand for the preservation of ancient monuments, and should any of the buried cities of Gobi ever be uncovered, no doubt they will be found to have been taken excellent care of. So, at least, it has been in the western plain of the Nile, where the sand is so fine as to be like a fluid, and has buried and preserved the monuments of Ipsambul so perfectly that not a feature is injured, nor are even the colours impaired.

The uses to which we put sand are many and various, but so familiar that the mere enumeration of them will suffice. The farmer and gardener use it for mixing with heavy clay soils, which would else be too stiff and air-tight for any crops to flourish in; the stone- and marble-cutter want it for sawing, grinding, and polishing, tor which last two purposes it is also used in many other trades, and the housewife scours her pots and pans with it, either in the natural state or in the form of sand-paper.

All this is obvious enough, and no one will doubt that sand is useful ; but who, unless he knew the fact, would guess that sand could also be made ornamental ?

Yet the inscription on an old German drinking-glass runs as follows : " I am beautifully clear and bright, and I am made of sand and ashes."

Very unpromising looking " dust " this, yet out of it come the crystal glass which sparkles on the dinner-table, and the window glass by which light is admitted to our rooms, as well as innumerable other varieties. How and when the art of glass-making was invented is unknown, but it was practised by the ancient Assyrians and Egyptians, as the specimens

found in the palace of Nimroud, among the relics of Babylon, and the tombs of Egypt, amply testify.

The discovery may well have been made by accident,
since lumps of impure glass are sometimes found among the
ashes when a rick of straw has been burnt down. For straw
contains, as has been said, a good deal of silica, and as it
also contains potash, the melting of the two together produces glass.

Many minute particles of glass are found in granite,
formed in a similar way, by the union of silica with the
potash or soda contained in the felspar.*

Most silicious sands—for it is of these only that we are
now speaking—are more or less yellow from the iron they
contain, and it is this which gives the green tinge to
common glass. The finest and whitest sand ever seen was
some brought from America and exhibited in the Crystal
Palace, in 1851, which was perfectly pure quartz and as
white as snow; but sand such as this is not easily obtained,
and, though millions of tons are scattered throughout the
British Isles, it is rarely found free from colouring matter
and other impurities, and needs washing and burning before
it can be used.

The chief places from which glass-making sand is obtained in England are Alum Bay, Lynn, Aylesbury, Ware-
ham, Reigate, and the New Forest ; and large quantities are
also brought from Fontainebleau, in France, as well as from

---

* Though the variety of felspar called orthoclase contains so much potash
as to be known as potash-felspar, no cheap and easy way has been found of
extracting the potash, which is therefore prepared from the ashes of plants.
Soda-felspar is called albite.

America, Australia, and New Zealand. But wherever it may come from and whatever may be its destiny, its past history has been much the same in all cases. Falling, in the first instance, from banks, cliffs, or mountain sides, it has undergone much grinding and pounding in the bed either ot glacier or river, and, after being washed, carefully sorted, and carried no one knows how many miles, has at last reached the resting-place, from which, after the lapse it may be of ages, it has been taken to the glass-house, whence it emerges in a totally changed form to begin a fresh career.

For, from the common wine bottle, made ot rough sand and other coarse materials, on through the many other varieties, till we come to the finest plate, crystal, and Venetian glass ; from the common ill-shaped tumbler to the exquisite ornaments fabricated by Salviati—every kind owes its existence, more or less, to the sand whose history we have been tracing.

But sand does not always remain sand until man happens to find a use for it. At the mouth of many rivers it is frequently found cemented into stone by the carbonate of lime brought down by them, and all the sandstone used for building was once nothing but loose sand lying along the shore, every grain of which has been more or less rounded by long-continued washing.

Some beds of sandstone are a thousand feet thick and more, and were, to all appearance, accumulated in bygone ages on an ancient shore which stretched from England across the North of Germany. It is not easy to say how the sand was consolidated, whether by pressure only, or by pressure and heat combined ; still less can we tell whence came the iron oxide which coats each separate grain of the red sandstone.

Sandstones consist mainly of grains of silica, which are generally intermixed with small particles of other minerals, and are cemented either by carbonate of lime, iron, or silica. Where both grains and cement are of silica, the sandstone would seem to have been formed by the agency of heat. Intensely hot steam, for instance, may have penetrated the mass of porous sand, partly melting each grain which, as it cooled, would be cemented to its neighbours.

The heat of molten lava would also have a similar effect, for the sand used to line furnaces is found at the end of a fortnight to be in part converted into a compact, close-grained stone, simply by the heat; and quartz rock, in which the grains can hardly be distinguished even with the aid of a microscope, was probably also formed by heat.

Pure quartz consists simply of silica, and crystallises in six-sided prisms. Cornish, Bristol, and other so-called " diamonds," are small bright colourless crystals of quartz, the purest variety of which is the rock-crystal used by spectacle makers, while the Scotch cairngorm, purple amethyst, chrysoprase, chalcedony, carnelian, onyx, heliotrope or blood-stone, and the precious opal, all consist of silica, variously coloured by other minerals and metals. We cannot, indeed, say that any of these stones have ever actually existed in the form of sand, but neither does there seem to be any reason why sand should not, in process of time, be transformed into any one of them.

But to return to the changes which we can see for ourselves to have taken place.

At the Cape of Good Hope there is a sandstone formation some 2,000 feet thick, which has evidently been affected in different degrees by heat, for in some parts it is stained red, brown, or yellow, by iron, in others it is perfectly white, as the red sand in a furnace becomes, and in others it is as compact as quartz.

There are in Great Britain some thirty well-known sandstone quarries of different qualities and colours, which supply large quantities of stone. Of ancient sandstone buildings in this country there are the abbeys of Tintern, Whitby, Rivaulx, the cathedrals of Ripon and Durham, and churches in Newcastle, Derby, Shrewsbury, Ludlow, and other places too numerous to specify. All these, and many others, some of them several hundred years old, are built of that which once was merely sand on the sea-shore, and as not a few of them have in great part crumbled away and are still crumbling, much of the stone will, probably, in time, return to the sea-shore again.

So much for the sand; and now for the mud, which, though carried farther out to sea before it is deposited, is seldom dropped more than two hundred miles from the coast. In a tideless sea or gulf it is deposited close in shore, or even at the mouth of the river, as witness the Nile delta, and the mud flats on the coast of Nova Scotia, which are derived from the neighbouring cliffs of shale and sandstone, the sediment being deposited tide after tide in layers, some of them as much as one-tenth of an inch thick, but generally much thinner.

The wide plains and low plateaux of Western Russia have evidently been similarly formed of mud and sand; but

I

as we travel eastward, and approach the volcanic rocks of
the Ural Mountains, both are altered by the heat, and pass
into schists and quartzites.   The muddy Russian plains are,
so geologists say, composed of some of the most ancient beds
of sediment in the world, and having been raised above the
waves ages ago, have never since been brought within their
reach or suffered much change of level.   In age they correspond with some of the Welsh rocks, but in Wales the mud
has been hardened and altered into slate, while in Russia it
remains pretty much in its pristine condition.

One of the most ancient heaps of mud in England is now
dignified by the name of the Longmynd, or Long Mount,
which is ten miles long, and even now, though it has lost
much of its height, rises more than 1,000 feet above Church
Stretton, and is not less than 26,000 feet thick.   Many of
the Longmynd rocks consist of innumerable thin layers,
some of them scarcely thicker than a sheet of paper, and
evidently once thin films of very fine mud and sand, though
now hardened into shales and slates of various tints, from
deep purple to grey and olive.

A gigantic mud-heap this is, certainly, and more ancient
than we can easily realise, even though we know that the
whole of England to the east of it, with its thousands of feet
of sandstones, coal-beds, limestones, chalk downs, &c., has
been formed since it was deposited.

The Longmynd rocks were originally deposited under
water in horizontal layers or films, such as those of the
Nova Scotia mud flats, the dark, gritty, coarse part of the
mud brought by each tide sinking first, and the finest
forming the top of each layer.   But, whatever other changes

may have taken place since those days, one is very evident: the great mass of mud has been upheaved until it stands almost on end, and, consequently, though it is quite plain that the layers were once in a horizontal position, they are now some of them almost vertical. They split off very readily along the lines of bedding, as one would expect them to do, for each layer of mud would naturally be slightly hardened before the succeeding one was deposited; but the fact of the rock thus splitting shows that, though of the same age and composition as some of the best

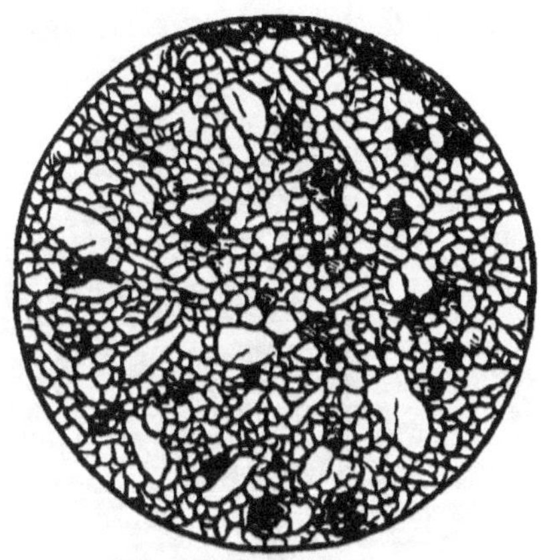

Fig. 22.—SLICE OF SHALE, SEEN UNDER THE MICROSCOPE AND HIGHLY MAGNIFIED.

roofing-slates in the world, it is, after all, not true slate, but shale (Fig. 22), for true slate splits, not along the lines of bedding, but at some angle, often at right angles, to them, and has clearly undergone some change which the shale has escaped.

Any fossils, such as shells, which occur in shales, are found lying unaltered in shape, with their flat sides parallel to the bedding, just as they would naturally have dropped through the water. When they occur in slate, on the other hand, they are distorted, and stand almost or quite on end, as they would do if the mud enclosing them had been subjected to great pressure from the sides.

A mass of clayey mud, containing grains of sand,

mica, &c., when sufficently pressed as to be reduced to half its original bulk, will be found capable of splitting into almost any number of thin layers at right angles to the direction in which the pressure is applied. It has, in fact, acquired the " slaty cleavage," and the various grains com-

Fig. 23.—SLICE OF SLATE, SEEN UNDER THE MICROSCOPE, AND HIGHLY MAGNIFIED.

posing it have ranged themselves with their flat sides facing the direction from which the pressure came. White wax, and even ice, may be made to acquire this cleavage in the same way by pressure, and it is also developed to a certain extent in biscuit by the mere application of the rolling-pin.

All mud which has been converted into true slate has, therefore, we conclude, been subjected to enormous pressure, and has thence acquired the property of easily splitting into thin plates, which makes it so valuable as a roofing material.

Fine mud is often rolled into roundish lumps and embedded in the coarser materials, and it is to these that are due the greasy whitish-green spots often seen in slates, the grain being too fine to bite the pencil. (Fig. 23.)

The sediment carried by the Rhine into the German Ocean consists chiefly of silica, alumina, and iron, and is just such as might hereafter form a clay slate, rich in iron.

The slate quarries of Carnarvonshire are the largest in the world, and give employment to 3,000 men ; the whole number employed in slate quarrying throughout Great Britain being about 15,000.

The "green slates" of Cumberland are composed of volcanic dust and ashes, which often contain large quantities of felspar, and thus form a very tenacious mud. Volcanic dust, converted into mud, has been found, more or less, wherever the bed of the ocean has been explored, vast quantities, as we have seen, are ejected during eruptions, and the lighter part is carried hither and thither by the wind. Dust from Mount Hecla has at times been conveyed to Denmark, and since much larger quantities have, no doubt, been dropped by the way, there are probably large accumulations in the German, as well as in the Indian, Ocean, which may be converted into shales or slates according to circumstances.

Ancient mud is found, however, in various other conditions besides ; some as soft clay, some mixed with lime, when it becomes marl, some as hard clay, and some so altered by heat as to be crystalline, and much harder than even the hardest and oldest of the Welsh slates.

Many of the most valuable clays occur in a semi-hardened state, and are blasted in rock-like masses ; but whether hard, or soft and sticky, all clays are essentially hydrous—*i.e.*, watery—silicates of alumina, and though always containing other minerals, are chiefly compounds of silicon oxide (silica) with alumina, the oxide of the silvery-looking metal called aluminium ; neither silicon, which is a black crystalline substance, nor aluminium, occur in the free state in Nature.   Clays owe their peculiar plasticity to the presence of combined water.*

The purest form in which silica is found is quartz, but even that contains a trace of alumina, and sometimes of iron as well; and the purest form of alumina is the blue sapphire, which is crystallised alumina, coloured by iron, and containing also a minute quantity of silica.   Neither silica nor alumina, therefore, is ever absolutely pure.

These two, with potash or soda, constitute the felspar which makes up so large a portion, not only of the granites, but of all the lavas, ancient and modern; and decayed felspar is simply *clay*, which varies in colour and consistency as it contains more or less silica, and more or less iron.

The purest clays—those used for the manufacture of porcelain—are more than half silica, and none are entirely free from iron, while many contain soda, potash, magnesia, &c.

The brick-making clays when burnt come out red, blue, or black, when they contain much iron, brown when they contain magnesia, white and dun-coloured when they contain lime.   More than a thousand million common bricks are made in England every year.

But Nature, too, has her potteries and brick-kilns, where she bakes her clay to bring out its colours, and one of the largest of these is in the "Bad Lands" of the Little

* Oxygen constitutes nearly half the weight of the earth, and is found combined with many metals and minerals. Soda, alumina, and lime are all compounds of oxygen, with the metals, sodium, aluminium, and calcium, none of which occur in nature except in union with oxygen.

Missouri, which are "bad" indeed for travellers, since the whole country is one huge labyrinth of ruins, which at times ook like those of some gigantic city (Fig. 24). Here are

Fig. 24.—BAD LANDS, NEAR FORT BRIDGER, IN THE UNITED STATES.

fragments of apparently Cyclopean masonry, walls, pinnacles, ramparts, terraces, obelisks, pyramids, fortifications, all heaped together in the wildest confusion, and baked of various colours from deep red and brown to pale yellow or china white. Here and there are towers and spires, looking as if they had just been freshly painted with vermilion, and reminding one of the churches of North Germany, and here again are piles of what look like crumbled bricks and mountains of potsherds, such as that of Monte Testaccio, near Rome, said to be composed of the broken pottery thrown away by the Romans.

Almost all the porcelain-clays are derived from the felspar of granite rocks, and usually contain spangles of mica. The soda or potash, having been attacked by the carbonic acid of air or water, is readily dissolved and washed away; then the silicate of alumina which remains is more slowly carried down the hill side in the form of fine powder, some of which is deposited in beds along the watercourses, while the finest and purest is carried into the valleys, or into rivers and lakes. The farther it is carried the more perfectly it is sorted, and hence the low-valley clays are often wonderfully fine.

The soil at the bottom of granite hills is, therefore, usually too stiff, while that on the top is too sandy, to be fertile.

The granite hills of Cornwall and Devonshire supply all the kaolin, or china-clay, used in the Staffordshire potteries. This is the finest and purest white clay known, and derives its name from Kaoling, "lofty ridge," the mountain from which it is obtained by the Chinese, who seem to have been the first to turn it to account. The place whence the largest quantity is derived in England is the neighbourhood of St. Austell, Cornwall, and when first raised it has the appearance and consistency of mortar.

China-stone, also used for making porcelain, is a product

of the same granite rocks, being simply felspar in a less advanced state of decay; and at Belleek, Fermanagh, kaolin is obtained from the undecomposed red granite of the district, which becomes white when calcined, the iron being extracted by magnets.

The Chinese began the manufacture of porcelain some two thousand or more years ago, and at the present day have large manufactories, King-ti-chin alone boasting, it is said, nearly 3,000 kilns. Chinese porcelain was first introduced into Europe in the sixteenth century, but no real progress was made in imitating it until the eighteenth century, when Böttcher, of Dresden, made first a red ware, and then white porcelain.

Earthenware of a coarse kind was manufactured in Staffordshire from a very early period.

In one of the ranges of the Appalachian mountains, known as Blue Ridge, the rocks (which are principally gneiss) are decomposed to a depth of fifty feet or more, and converted into a reddish, greasy, brick-clay; for gneiss is composed of the same three minerals as granite, but these are arranged in plates instead of grains, and in the Blue. Ridge rocks the plates of quartz may be seen in their original positions, embedded in clay. In some of these rocks the various stages of decomposition may be well observed. Thus the upper part is completely kaolinised, and almost entirely freed from the iron which gives the red tint to the coarser clay; lower down the rock is partly decomposed, and lower still quite unchanged, so that it seems evidently to have been decayed from without, probably by water charged with carbonic acid filtering through from the surface. The iron oxide seems then to have been dissolved out and carried away to the foot of the range, where there are immense deposits of iron ore.*

West of Blue Ridge alumina is found in considerable quantities, combined with a small amount of iron and silica ; but instead of being in the soft, sticky state of clay, it is intensely hard, being in fact crystallised. Were it not that the crystals are all flawed, and therefore valueless as gems, they might almost be called a mass of sapphires and rubies, for many of them are coloured with the most beautiful tints of pink, blue, and deep ruby. As it is, the mineral is called corundum, and is most useful for grinding and polishing, and, being much harder than emery, and prepared at much less cost, is likely to be a formidable rival. Emery itself is but another form of alumina, containing a large admixture of iron, and is obtained chiefly from the decayed rocks at Cape Emeri, in the island of Naxos.†

The opaque stone, known as the Oriental turquoise, is a phosphate of alumina coloured by copper.

And thus from clay and mud we have come round to

* The gneiss and granite rocks of Brazil are similarly decomposed to the depth of 100 feet.

† The hardness of the sapphire being 100, that of corundum is 77, and the emery of Naxos, 46. By melting china clay with red lead, the silica is extracted, and after exposure for several weeks to intense heat, the mixture, on being allowed to cool, is found separated into two layers, the upper, mainly silicate of lead and glassy-looking, the lower, crystalline, and containing perfect, but colourless, crystals of alumina, which are specimens of the corundum, and when coloured by the addition of iron, cobalt, &c., differ in no respect in composition or appearance from the Oriental ruby and sapphire.

precious stones again. Whether the latter have actually been formed out of the decomposed materials of older rocks or no, there is apparently no reason in the nature of things why they should not have been, for in the Blue Ridge rocks we seem to have almost the whole process before us. There is the gneiss, the decayed felspar, greasy brick-clay, kaolin, and finally the corundum, which seems to have needed only slightly different treatment in the great laboratory to convert it into precious stones.*

However this may be, we have traced our sand and mud far enough for the present purpose, and have seen how Nature has sorted and transformed them into sandstones, clays, shales, slates, &c., and how man has taken them up where she left them and has used them from the earliest ages to pile up those vast heaps of stone, brick, glass, tiles, and slates, which we call towns and cities.

What was the Tower of Babel but baked mud? And what is the "modern Babylon" in the main, but a vast transformed mud-heap, rivalling the Longmynd in size, though certainly not in beauty?

---

* The old naturalists said heat was required to ripen precious stones. Certainly they are found only in the south ; though the common topaz is found by the hundredweight at Falun, and crystals of common emerald, several feet long, occur in the felspar quarries of Finland

CHAPTER IX.

WHAT BECOMES OF THE "DUST"—CORAL ISLANDS, ETC.

What becomes of Nature's "Dust"—The Invisible made Visible—Rivers supply the Inhabitants of the Ocean—Oyster-shells, Mother-of-Pearl, and Pearls—The Largest existing Shell—Coral Polyps, their Stony Skeletons, Coral Reefs, Coral Islands—Depths of the Ocean—Microscopic Shells—Stone Lilies--Limestone and Marble.

SAND and mud do not by any means represent all that is carried down into that universal receptacle, the great ocean, which is receiving fresh additions every moment of the day and night, and yet never gets overfull. The clearest of clear streams is as surely conveying something to it as the mud-laden river; and as everything that water can dissolve reaches the sea at last, sea-water contains at least a trace of every soluble mineral and metal in the world.

Roughly speaking, however, the average composition of 1,000 grains of sea-water is as follows:—

Water	962·0
Sodium chloride *	27·1
Magnesium chloride	5·4
Potassium chloride	0·4
Bromide of magnesia	0·1
Sulphate of magnesia	1·2
Sulphate of lime	0·8
Carbonate of lime	0·1

* Common salt.

This at first sight appears remarkable, for it will be remembered that carbonate of lime, of which this analysis shows so small a proportion, is the mineral carried down in the largest quantities, at least, by the European rivers*; and that the chlorides, though they seldom fail altogether in any river or stream, are yet conveyed in comparatively minute quantities, except by a few small rivers which are especially rich in them.

The sun, when he drinks, takes pure water, leaving salt, lime, and all else, behind in the ocean ; what then becomes of the enormous quantity of carbonate of lime which the Rhine, Rhone, Loire, and Thames, to mention no others, are continually pouring into it ?

It is in the ocean still, but it has changed its form ; or rather has again acquired a form and has once more become visible.

The rivers gather materials from all parts of the earth, and their work is not for nothing; they cater on land for the inhabitants of the sea, and what they pour in is all wanted there, and the reason why there is so little dissolved lime in sea-water is that it is required for so many purposes.

Where, for instance, would the oyster get the material for its shells, if the rivers did not supply its need ; for, being rooted to its bed, all it wants must be brought within its reach ? Moreover, the lime must be dissolved before it can be used ; the oyster would not be able to make anything of a lump of chalk or limestone ; for, like all other shell-covered mollusks, it has to swallow the materials of which its habitation is made. These are secreted by a sort of

* Chap. IV.

transparent skin, or membrane, called the "mantle," which covers the body more or less, and shapes and colours the shell according to its own peculiarities.

The Rhine alone carries to the sea every year carbonate of lime enough, according to Bischof, to make the shells of 332,539,000,000 oysters, and as it empties itself into the German Ocean, no doubt the oyster-beds on the coast of Kent are supplied with their house-building materials in part at least by its means—materials which have been collected for the purpose far away in Germany and Switzerland.

Oysters are full grown in about four years, and it is said that in that time, in order to obtain enough material, ten of them must swallow from 345 to 587 lbs. of sea-water, or from 5·2 to 8·9 cubic feet; but this is supposing they would extract the whole of the lime, and as that is unlikely, they must actually swallow much more.

The artificial beds at Whitstable and Faversham, in Kent, alone extend over nearly twenty-seven square miles, and the natural oyster-beds in America, some of them, cover a million acres; but it is quite beyond the power of figures to express the number of oysters contained in such beds, for it takes 1,600 to fill a bushel measure. One oyster might, it is said, have offspring enough to fill 12,000 barrels.*

Numerous beyond counting, however, as are the oysters, mussels, whelks, and periwinkles, they are mentioned first, not as being the chief consumers of lime, but because they

* Five million dollars' worth are annually consumed in New York alone.

are such familiar objects, to all Londoners at least; and they acquire fresh interest when it is remembered that every particle of the material for their shells was brought to them invisibly, perhaps from the Alps, perhaps from the English hills and downs, and perhaps from countries far away.

Shells are of two kinds; some, like the foreign cowries, are hard and compact as porcelain, while others, like the oyster, are formed in layers, and are often covered with a skin; but all consist chiefly of carbonate of lime and gelatine (the hard sorts containing much less gelatine) with varying proportions of sulphate of lime, phosphate of lime, carbonate of magnesia, and iron. Pearls, which are secreted by some other bivalves, as well as the oyster and mussel, consist also chiefly of carbonate of lime, and are, in fact, of the same composition as the mother-of-pearl forming the shell. This is arranged in concentric layers like the coats of an onion, round some minute particle of refuse, such as a grain of sand, which the animal has been unable to remove from its dwelling, and therefore covered up as the only way of diminishing the annoyance.

It is needless to specify the various uses to which mother-of-pearl is put, from the making of shirt-buttons, to the inlaying of *papier mâché*; but whatever its quality and colour, its history is always much the same, and so, too, is that of the shell cameos, though their texture differs from that of mother-of-pearl.

Cameos are cut from univalves—shells formed all in one piece, like the cowry, limpet, periwinkle, &c., and consisting of three layers of calcareous matter, which in some are all of different colours. In 1875, 300 persons were employed in

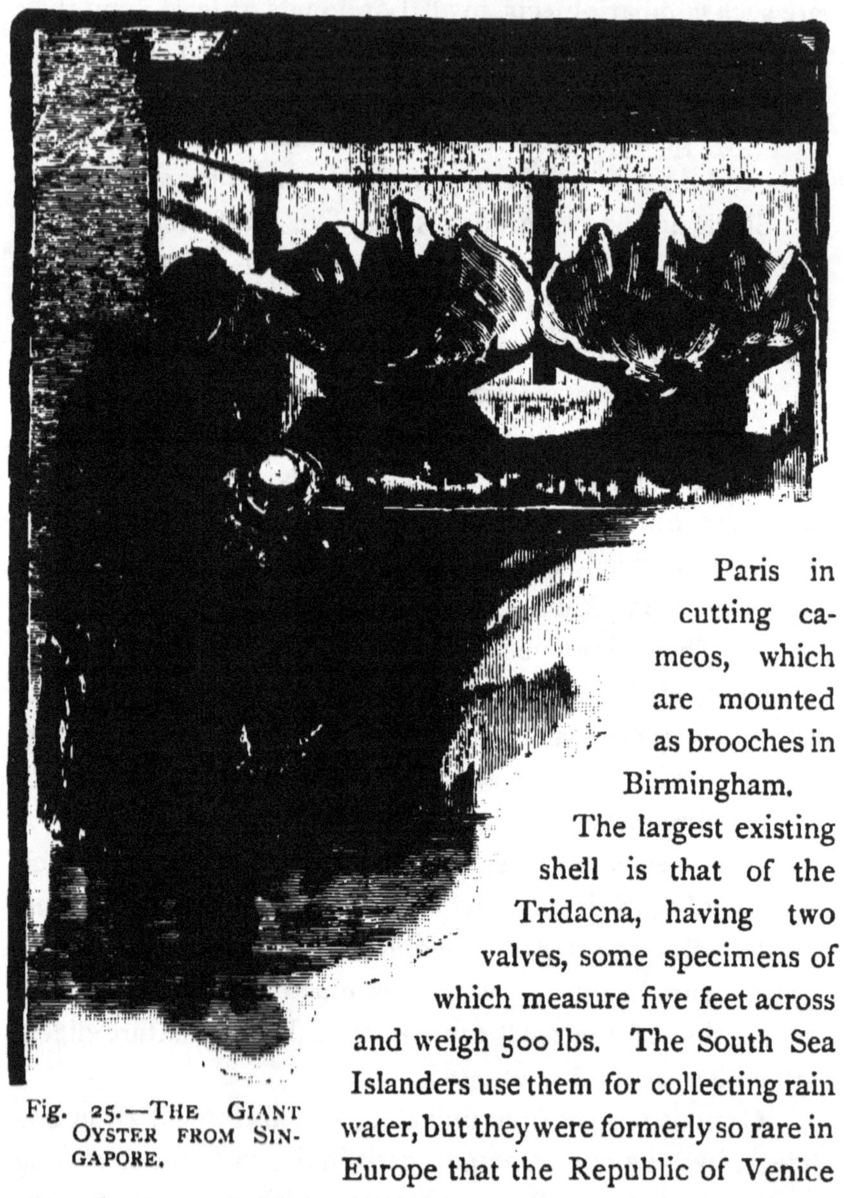

Fig. 25.—THE GIANT OYSTER FROM SINGAPORE.

Paris in cutting cameos, which are mounted as brooches in Birmingham.

The largest existing shell is that of the Tridacna, having two valves, some specimens of which measure five feet across and weigh 500 lbs. The South Sea Islanders use them for collecting rain water, but they were formerly so rare in Europe that the Republic of Venice thought one valuable enough to be presented to Francis I., who gave it to the Church of St. Sulpice, where it may still be holding the "holy" water. The shells are more common now, and are often employed of all sizes for this purpose. One from Singapore (Fig. 25) was shown in the London International Fisheries Exhibition, in 1883. Its weight was 3 cwts., 3 qrs., 14 lbs. ; its length, 3 ft. 4 in. ; and its breadth, 2 ft. 2 in.

In some parts of the Arctic Ocean the bed is covered with "urchins" or sea-eggs, and these, as well as the innumerable crabs, lobsters, cray-fishes, and many others, all require lime for their shells or armour.

Passing over these, however, we come to other animals, which though individually much smaller, yet take up far more space than any beds of oysters or other mollusks, and in fact cover an area so vast as hardly to be estimated at all. We mean, of course, the coral-polyps, "coral *insects*," as they are erroneously called in certain pieces of poetry. They are, however, no more "insects" than the oysters ; and in spite of the same poetry, they cannot accurately be described as "builders" or "architects," nor held up as examples of industry.

No one thinks of speaking of the grass in the meadow as "toiling" to make hay, nor does one praise baby for his industry because he has grown since last year, and coral-making is no more toilsome than bone-making. What we call "coral" is in fact the animal's skeleton, not its house.

The coral-polyp, as the creature is called, in allusion to its many arms or rays, much resembles the garden-aster in appearance, and is still more like the sea-anemone, which indeed gives one a better idea of it than any description can do.

J

Both animals consist of a disk and bag. The former, which is often brightly and beautifully coloured, is set round with fringed rays ; these are in reality hollow tubes, which close over all the prey brought within their reach, and convey it to the mouth, which is placed in the centre of the disk, the stomach being contained in the bag beneath. Thus far the anemone and the coral-polyp are alike, though the one may be many times larger than the other ; but if the anemone's bag were cut across, we should see that it was composed of several divisions, and it would present much the same appearance as an orange or lemon when similarly cut. In the anemone, as in the lemon, these partitions are formed by a skin or membrane, but in the coral-polyp they are hard and solid, and consist of carbonate of lime.

Some corals are, to all appearance, true sea-anemones, their stony skeletons being so completely hidden as to be un-suspected by those not in the secret ; but there are others which, instead of living singly as these do, grow together in groups of thousands and hundreds of thousands, each tiny individual having its own flower-like disk, mouth, and stomach, but being at the same time as closely united with the rest as the twig is with the tree.* The whole mass is indeed one, but fed by many mouths, and the various in-dividuals bud, and branch, and grow in such a truly plant-like way that their name of zoophytes (animal-plants) is most

---

* The red coral grows up in a sort of branched stem, every branch being terminated by an open-mouthed polyp. The skeleton, which belongs to the whole colony in common, is covered by a soft body by which it is deposited. In the white coral, besides the common skeleton, there is a special one for each individual. This is shaped like a cup and divided by radiating parti-tions, and the cups are united into a common branch.—*Huxley.*

appropriate. In shape they imitate almost every plant which grows on the land. There are branching trees, from six to eight feet high, covered with starry polyp blossoms; there are shrubs of various shapes, tufts of imitation rushes, pinks, feathery mosses, broad leaves studded with daisy-like flowers, cacti, fungi, and lichens, in endless variety of

Fig. 26.—ONE OF THE ASTRÆA CORALS (*Faria pallida*).

beauty. Some colonies grow in the shape of graceful vases, which measure three or four feet across, and are composed of sprigs and branches representing countless multitudes of individuals. Others again, as Astræa, shape themselves by common consent into solid domes, with a diameter of from ten to twenty feet, which are scattered all over with stars of purple or emerald-green (Fig. 26). These immense groups all spring from one germ, and are so intimately connected that an injury to one individual is felt by all the rest, which at once close their flowers. Yet a twig of coral may be broken off without being killed; in two or three hours it will recover sufficiently to open out again, and if placed in a suitable position will soon begin to sprout and grow after the family pattern.

It has not at present been possible to calculate the rate of growth with much certainty, but twenty corals planted on a sand-bank east of Madagascar, grew nearly three feet in height and several feet in length in the course of six or seven months; and the copper sheathing of a vessel which had spent twenty months in the Persian Gulf, was covered with a hard crust of coral two feet in thickness. No doubt, however, the rate of growth varies in different species.

Most of the Bermuda corals resemble anemones or groups of anemones, the stony skeleton being entirely concealed; but in a few, as the brain-coral, which grows in the shape of large domes, it is only just coated with a film of yellow or greyish living matter.

Like the bog-mosses, many of the coral-polyps go on growing above while they die below, and thus the large solid domes of Astræa are quite dead within, the living portion being confined to the surface and only half an inch thick, or even less.

The great masses of coral rock, called reefs, which in some places extend for hundreds of miles along the coast and form natural breakwaters, are composed only in part of living coral, and consist chiefly of the consolidated *débris* of many kinds of coral and corallines, the shells of mollusks, and the tubes of many sea-worms.

The great reef on the N.E. of Australia is 1,200 miles

long, and seventy miles wide near the southern end, while that on the west of New Caledonia is 400 miles long.

The Bell Rock lighthouse, in the German Ocean, though 112 feet high, is often completely buried in foam and spray during a ground-swell, though there be no wind to lash the waves. But the great rollers of the Pacific are far heavier than any waves in the German Ocean, and during a storm they do as much damage in the coral groves as a gale does among the trees on shore. Sometimes, especially where the reef has been weakened by boring mollusks, masses twenty and thirty feet long are torn off, many a coral tree is prostrated, boughs and twigs are torn off, and the polyps over a large surface destroyed. The dead parts are, however, soon overgrown and protected by smaller encrusting corals and other zoophytes, as well as by serpulæ, mollusks, and lichen-like nullipores, just as the dead trunk of a tree is overgrown with mosses, &c.

There are deep-sea corals, but the reef-making kinds do not flourish at a greater depth than from 120 to 150 feet, and prefer being within reach of the light, but where they may be either quite covered or constantly washed by the waves, since a very short exposure to the sun kills them. Only two species, however, are able to stand the full violence of the breakers on the upper and outer edge of the reef, and of these one grows in thick vertical plates, while the other grows in masses from four to eight feet broad and nearly as thick. These massive kinds actually thrive best where they are most exposed, and are less perfect in sheltered spots, where more delicate kinds flourish.

The largest number of species, however, are found where

the heat is intense and the water calm, as in the Red Sea, which has some 120 varieties. No coral can live except in quite clear water; and as all rivers bring down more or less sediment, gaps usually occur in the reef opposite their mouths.

Since reef-corals cannot live below a certain depth, it is quite evident that they have not worked their way up from the bed of the ocean, as used to be believed, and the method in which the numerous coral islands, or atolls, which stud the Pacific have been formed, seems to be this:—

In the first place the coral has grown upon the submerged rocky bed immediately surrounding some island, and has gradually formed a fringing reef. Here it grows upwards until it gets too near the surface, when it will cease growing in that direction and grow outwards instead, the inner parts dying, being broken into fragments, ground up into sand, and consolidated into a compact rock, and fresh shoots constantly growing in their place.

As long as the island remains stationary, the reef can do nothing but grow outwards, but if the island should sink a little, the sea will flow in and a channel will be formed between the island and the reef, which is now called a barrier reef. This apparently is what has taken place on the coast of Australia, where the channel between it and the reef is generally from five to fifteen, but, in one part, nearly a hundred miles wide.*

But supposing that the land should continue to sink until

---

* Mr. Wallace tells us that the outer side of the Great Barrier reef sinks 2,000 feet, which shows how much the rock upon which the corals first grew must have subsided.

nothing of it remains above water, except, perhaps, what was once the top of some hill or mountain, and that after a while this, too, disappears; then nothing will remain to mark the spot where the island was but the reef which once encircled it. For the coral will have continued to grow upwards as its foundations sank. Had there been a sudden plunge into the depths below, it would have been killed; but this slow, gentle subsidence, so far from being injurious, merely gives it the necessary room to grow, and the deeper the foundation sinks, the thicker the reef will be.

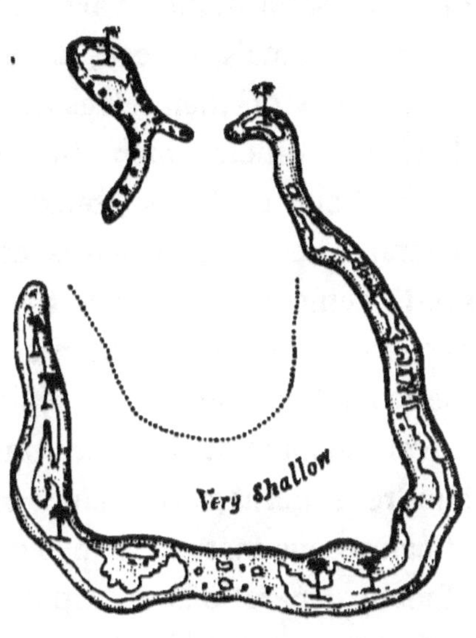

Fig. 27.—GROUND PLAN OF KEEL-ING ATOLL. (*After Darwin.*)

The whole of the Pacific Ocean is scattered with atolls (Fig. 27), or rings of coral, some of them many leagues across; and all, larger or smaller, are so many memorial stones, marking the site of a buried island. The reef is below water, of course, but its position is clearly marked by the line of snow-white breakers, which are incessantly dashing themselves against it. By their means heaps of coral sand are gradually formed and piled upon the reef, forming little islets which are heaped higher and higher, converted into dry land, and covered with vegetation. It is not often, however, that the whole atoll is raised above the water; generally speaking it is only here and there that it rises into an islet crowned with feathery cocoa-palms, the rest of its outline being marked by the snow-white breakers.

The Laccadives, or "lac of islands," and the Maldives, or "thousand islands," are just a series of such atolls.*

Few animals, indeed, have left such vast and enduring monuments of themselves as these humble zoophytes; but there are others which must not be passed over. Outside the harbour of Pernambuco there is a reef of sandstone, several miles long, which is composed of grains of silicious sand, cemented together by carbonate of lime. Though exposed to all the violence of the great Atlantic waves, with their load of sand and sediment, which are unceasingly driven against it by the trade-wind, the bar has lasted hundreds, perhaps thousands of years, thanks to a few inches of calcareous matter, formed chiefly by the growth and decay of many generations of serpulæ, together with a few barnacles and some paper-like layers of a sort of sea-lichen, called nullipora. The inner surface of the sand-bar, which has no such protecting coat, is visibly worn in spite of its sheltered position.

Off the Bermudas there are reefs which are mainly composed of these serpulæ, whose tubes one may often see on oyster-shells, though this gives one no idea of their real

* It has been suggested that, as the bed of the ocean is not a plain, but diversified by hills and even mountains, and as enormous quantities of shells are constantly falling to the bottom, some of these hills may be so heightened as to reach the point at which deep-sea corals, annelids, sponges, mollusks, &c., can live, after which they would increase more rapidly until they reached the zone of the reef-corals, which would grow upwards as long as they could, and then outwards.

beauty, for when alive, the little inhabitant waves a plume of brilliantly-coloured feathers—its breathing apparatus—from the door of its home.

The nullipora is one of the numerous sea-plants, like the corallines, which take up large quantities of carbonate of lime, and though not hard, as the latter are, is very tough. Both grow in most luxuriant profusion among the Bermuda reefs, and are of great beauty. As we know it, the coralline, with its hard, jointed stems, is either pale pink or lilac, or more often bleached to the whiteness of bone by sun and wind; but in the Bermudas the prevailing colour is green, which varies from a bluish to a grass-green tint, with here and there a tuft of plum colour, and white is rare. One species of nullipore is of a brilliant peach colour, and has thin, stiff, moss-like branches, the tips only of which are alive, and another grows in lichen-like sheets.

Both plants take up so much lime, that they are capable of forming masses of calcareous matter two or three feet thick by their successive growth and decay.

It had been supposed, until within the last few years, that the depths of the ocean were devoid of both animal and vegetable life, but the voyage of the *Challenger* has shown that this is a mistake. Only an infinitesimal portion of the ocean floor, at depths over 2,500 fathoms,* has even now been explored, we must remember; but so far as this has been done, animal life is found everywhere— though it is less plentiful in extreme depths—while plants, though stragglers may be found here and there, are practically limited to depths under 100 fathoms. Sponges exist

---

* Three miles.

everywhere, though less abundantly beyond 1,000 fathoms, and one coral lives at all known depths.

The whole of the Atlantic basin, to the depth of about 2,200 fathoms, is covered with an almost uniform greyish sediment, which to the naked eye appears like mud, but as seen by the microscope consists of very minute shells and fragments of shells. The upper surface of this "ooze" is of a creamy consistency and is made up chiefly of whole shells, with fragments of sponge spicules, and a considerable number of the larger shells of dead mollusks, more or less broken and worn. At all moderate depths, sponges, corals, star-fishes, &c., live in and upon this ooze, The next inch or two below the surface is of firmer consistency, the tiny shells being more or less broken up and cemented into a calcareous paste, and beneath this, again, whole shells and even fragments are rare, and the paste is almost uniform.

What thickness this sediment may have attained is unknown; but in past ages hundreds of square miles have been covered to the depth, in some places, of a thousand feet or more, by a deposit evidently also formed at the bottom of the ocean, and bearing a very strong resemblance to the Atlantic ooze, which may fairly be called chalk-mud—chalk which has not yet had time to harden into rock, but which, should it do so hereafter, will greatly resemble the chalk of the English downs and cliffs. For that, too, consists mainly of microscopic shells, which, though almost or altogether invisible to the naked eye, are of shapes as various and beautiful as any of the larger ones, some of which they closely resemble. (Fig. 28.)

The shells of these foraminifera, though so minute that 3,800,000 have been counted in an ounce of sand from the Antilles, are divided, like those of the nautilus, into chambers connected by minute openings, from which they take their name.*   The shells of some species are also perforated with innumerable tiny holes through which the inhabitant protrudes hair-like filaments of jelly.   Though a few are found

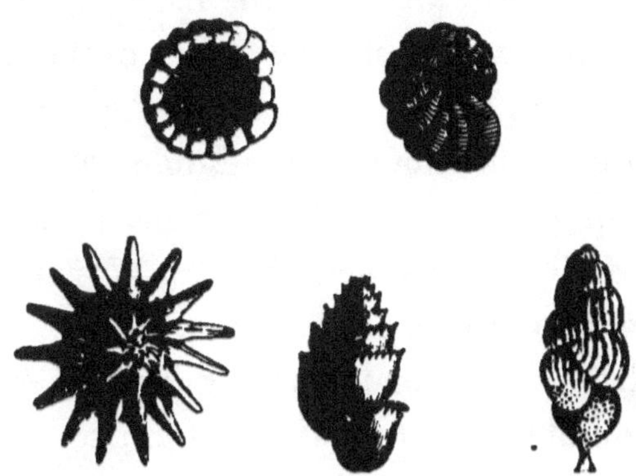

Fig. 28.—FORAMINIFERAL SHELLS.

alive in the ooze, that is not their natural home, for they live on the surface of the ocean, and if they are as numerous to the depth of 100 fathoms as they are in the track of the tow-net, every square mile of the ocean must contain sixteen tons of carbonate of lime, in the form of calcareous shells.   These, as the inhabitants die, fall down in a constant rain, and accumulate on the ocean-bed.   In some cases, half the sand on the sea-shore is made up of them, and they have had a large share also in the formation of coral islands.

They abound to such an extent in the limestone used for

* Latin, *foramen*, a hole.

building in Paris, that that city, as well as many towns and villages in the vicinity, may be said to be in great part built of them. They are also very numerous in the chalk which extends from Paris to Tours, a distance of fifty miles. One species alone has formed enormous beds in Russia, while the stone of the Great Pyramid is composed chiefly of others called Nummulites,* which are the most highly organised of all the foraminifera, and are the giants of the race, being of about the size and shape of a shilling (Fig. 29).

One band of nummulitic limestone, often 1,800 miles broad and in many parts of enormous thickness, extends along the Mediterranean and through Western Asia to the North of India and China.

But though occurring in large quantities in most limestone, foraminifera do not usually form so large a proportion of it as they do of chalk, which may well be called "foraminiferal limestone." †

A slice of ordinary limestone, ground so thin as to be transparent, when examined by the microscope, shows that it is made up of all sorts of minute fragments—bits of shell, seaweed-skeletons, and coral, with perhaps a few perfect microscopic shells as well. It bears, in fact, a strong resemblance to the coral-reef rock, which contains *on the whole* but few perfect remains, owing to the vigorous way in which it is pounded into sand by the waves, though large mollusks, and even the huge Tridacna, do sink or burrow their way into it and are preserved.

* Latin, *nummulus*, a small piece of money.

† Shells of foraminifera may often be found in the sand which comes from a new piece of sponge.

Some of the great Swiss mountains may be just parts of an ancient coral-reef, for though the Jura is some thousands of feet thick, it is but three-quarters of the length of the barrier-reef now existing off the coast of Australia; and

FIG. 29. — NUMMULITIC ROCK, SHOWING SEVERAL SPECIES OF FORAMINIFERA.

although at the present day coral does not grow anywhere north of the Bermudas, it evidently did so in the warmer seas of ancient times, for many of our Devonshire and Bristol marbles are almost entirely composed of fossil corals.

Though the name of "marble," is often inaccurately applied to any stone which will take a polish, it really belongs only to hardened limestones, many of which owe their beauty to the fossil remains which they enclose.   Some consist almost entirely of sea-shells, others, which occur only in small beds, of small fresh-water snail-shells, while another, again, is crowded with ammonites, which somewhat resemble the modern nautilus, and also with the internal " shell," or, rather, single bone, which formed the skeleton of the ancient cuttle-fish.   Perhaps the most beautiful marble of all is that found in large quantities in Derbyshire, which is composed of the remains of encrinites, or stone-lilies, a sort of stalked star-fish, having cup-shaped bodies and fringed rays with numerous joints which they could spread out to entrap their prey.   The stem was formed of innumerable small, star-shaped pieces of hard carbonate of lime, and was so pliant as to bend to and fro before the waves.   The upper part of one of these lilies is said to have consisted of nearly 27,000 separate joints.

Deep beds of encrinital limestone have been formed of the skeletons of these lilies, which, though sometimes found almost entire, are generally broken into a thousand fragments.   It had been thought that very few of these stone-lilies now existed, but numerous specimens have lately been found as far north as Siberia.   In ancient times, however, they were far more abundant, and in the north of Europe and America there are vast strata, composed entirely of their remains.

It is no mystery, therefore, what becomes of the mountains of lime which are carried into the ocean, where the innumerable hosts of mollusks, the still larger hosts of foraminifera and coral-polyps, as well as the corallines and other sea-plants are waiting for it. In the West Indian seas, where coral islands abound, it is disposed of so rapidly that the water contains less lime than elsewhere.

And now as to the way in which these animal and vegetable remains may be converted into rock and marble.

The first part of the process may be well observed in the Bermuda Islands, which are surrounded for twenty miles by fine coral sand which the waves have ground from the coral reefs. Some of this, no doubt, is washed into the cracks and crevices of the reef, and helps to consolidate it, and some may be re-dissolved. But much of it is washed on shore, dried by the wind, and blown into hills forty or fifty feet high, which are driven farther and farther inland, and unless their progress is arrested by the planting of shrubs, &c., often overwhelm both gardens and houses (Fig. 11).

Exposure to the air deprives the sand of the animal matter mixed with it, and then, like any other carbonate of lime, it is readily dissolved by water and carbonic acid. When rain falls, therefore, a little of the lime is taken up, sinks with it through the sand, and, as the water evaporates, is deposited as a cement which binds the loose grains together. This process being constantly repeated, at last converts the sand into rock of various degrees of hardness, some being so compact as to be almost like marble, and capable of taking a fair polish.

All the sand is not equally affected, for the rain seems frequently to follow certain particular channels which it keeps open and hardens, and this, together with the fact that all the lime dissolved is not re-deposited, gives rise to the wonderful caverns already described, whose contents have of course been carried back to the ocean. Thus the shells and skeletons of countless former generations are again converted into shells and skeletons by their modern descendants and representatives.

On the west of Ascension Island, to take one of many similar instances, the beaches are heaped with rounded fragments of shells, corals, &c., which, though loose on the surface, are, at the depth of a few feet, cemented into solid stone, some of which is actually too hard for building purposes and has the ring of flint. It contains very few perfect shells, but each rounded fragment may be distinctly seen to be surrounded by a husk of transparent carbonate of lime, and the stone is nearly as compact as marble which has been subjected to heat and pressure.

For this is the final stage in the process by which coral- or shell-sand is crystallised at last into the sparkling marble, much resembling loaf-sugar, which the sculptor uses for his statues. Not a trace of shell or other organic remains is now left, and the stone is so fine-grained and snowy-white that we could never guess its origin, did we not in many instances actually see what has taken place. For where ordinary limestone, whose composition we can see, has come in contact with a stream of lava, all traces of its original structure have disappeared. It has, in fact, been melted, and on cooling down has become perfectly crystalline; a little farther off the stone is hardened and partially crystallised, and farther off still it remains unaltered, being beyond the reach of the heat.

Crystallisation is a purifying as well as hardening process, and the Carrara marble, which is sent from the Apennines to all parts of the world, is found in large masses of dazzling whiteness, embedded in stone of a very dark colour, consisting of various impurities, such as animal and vegetable matter, particles of flint, &c., which have been driven to the surface by the process of crystallisation. Part of this dark stone is usually left on the blocks of marble as a proof of its quality.

The most highly transparent crystals of carbonate of lime are those known as Iceland spar, which is as clear as glass and quite colourless.

Some limestones, such as those of Westphalia, contain a large amount of iron which, though probably derived from sea-water, is more than could be separated from it either by animals or vegetables, though both take up small quantities, and it is to iron that red coral owes its colour. The bulk of the limestone-iron was, however, probably separated by a purely chemical process, being deposited by the water in the place of a corresponding quantity of carbonate of lime.

So, too, with the magnesian limestones, called dolomite, which contain, some of them, more than thirty per cent. of carbonate of magnesia, while fresh coral contains hardly so much as one. It has been suggested that, as the ash of certain fresh-water plants contains a good deal of magnesia, some dolomites may have been formed by them, just as it is thought that some dark-coloured, carbonaceous limestones may have been formed by the decay of different species of Chara, a lime-absorbing plant, which grows in great profusion in some lakes of North Germany.

K

But some dolomites are evidently of coral-reef origin, and since we know of no plant or animal which could have taken up magnesia in these large quantities, we must conclude that it was deposited directly from the water.

Wherever the depth of the Atlantic exceeds 2,200 fathoms, there the calcareous ooze passes gradually into a very fine reddish or chocolate-coloured clay, derived apparently from cosmic and volcanic dust, and the decay of pumice-stone, which floats long distances before it becomes so water-logged as to sink. The clay contains a few foraminifera, &c., but as these exist everywhere, from the equator to the regions of polar ice, and their shells must be falling in ceaseless showers all over the oceans, one wonders what has become of them, as well as of the shells of those mollusks which float about in multitudes in mid-Atlantic.

The explanation seems to be that, owing to the greater depth of water they have to sink through, and the larger amount of carbonic acid present at great depths, they are dissolved before they reach the bottom, and the shells of the mollusks dissolve more quickly than those of the tiny foraminifera.

Quickly, however, they none of them dissolve, for the office of these and other marine animals is just this: to separate the lime from sea-water, and *protect it from solution* by combining it with organic matter, and they succeed so well that their calcareous structures dissolve with difficulty, even in hydrochloric acid.

CHAPTER X.

WHAT BECOMES OF THE "DUST"—FLINT, SALT, ETC.

What becomes of Nature's "Dust"—Difference between Atlantic Ooze and
Chalk—Origin of Flints—Sponges—An Extensive Pumping Apparatus
—Microscopic Population of the Ocean; Rapid Multiplication—Micro-
scopic Plants, Seaweeds, Kelp-making—Where the Salt of the Sea
comes from—Animals killed by drinking River-water—The Great
Salt Lake—What would happen if the Straits of Gibraltar were closed
—The Dead Sea—Circulation of the Ocean.

WE mentioned in the last chapter the strong resemblance existing between the Atlantic ooze and
the chalk, which, wherever found, has evidently been formed
in a similar way. But there is one great difference between
them.

Chalk, especially that of the English cliffs, consists
almost entirely of carbonate of lime, and such small quantities of other minerals as it contains are as equally mixed
with it as if the whole had been well stirred. Flints are
very commonly associated with chalk, and sometimes occur
in layers or sheets at almost regular intervals, but not a
particle of flint is scattered about in the chalk itself.

The Atlantic mud, on the other hand, contains no
flints, but does contain a large proportion of silica (twenty
to thirty per cent.), in the shape of cases, sheaths, and
skeletons of minute animals and vegetables, and the bed
of the Pacific is to a still larger extent covered with these
silicious remains.

Silica is not enumerated among the minerals contained in sea-water, because in 1,000 grains the quantity is so small as to be little more than a trace. Silver, gold, and many other metals and minerals, are omitted for the same reason. But if there be but a trace in 1,000 grains the amount dissolved in the whole of the ocean will be far from trifling. It has been calculated, for instance, that the ocean must contain some two hundred million tons of silver.

With regard to the silica, we know that it exists dissolved in all water more or less, and is, therefore, conveyed to the ocean in large quantities; but, like the lime, it is wanted by such innumerable living things that little remains in the water.

Silica also is greatly attracted by dead and decaying substances, and collects round dead sponges, bits of coral, shells, &c., and sea-urchins are frequently found filled with flint, as well as embedded in it.

So many flints have more or less the shape of sponges that it seems probable these formed the nucleus round which the silica collected; and, indeed, when a thin slice of flint is magnified, it is very often found to contain various minute organisms, such as sponges are known to feed upon.

Toilet sponges consist of fine elastic fibres, resembling silk in composition, and closely woven together. But these form only the skeleton, and when alive were covered within and without with a film of jelly, like white of egg, but consisting of numerous individual animals or living cells, which were as perpetually changing their shape as a drop of liquid kept in constant motion.

Some of the cells being provided with hair-like lashes (*cilia*), which they keep waving to and fro, draw the water into the innermost recesses of the sponge through the minute pores, and expel it again through the large holes,

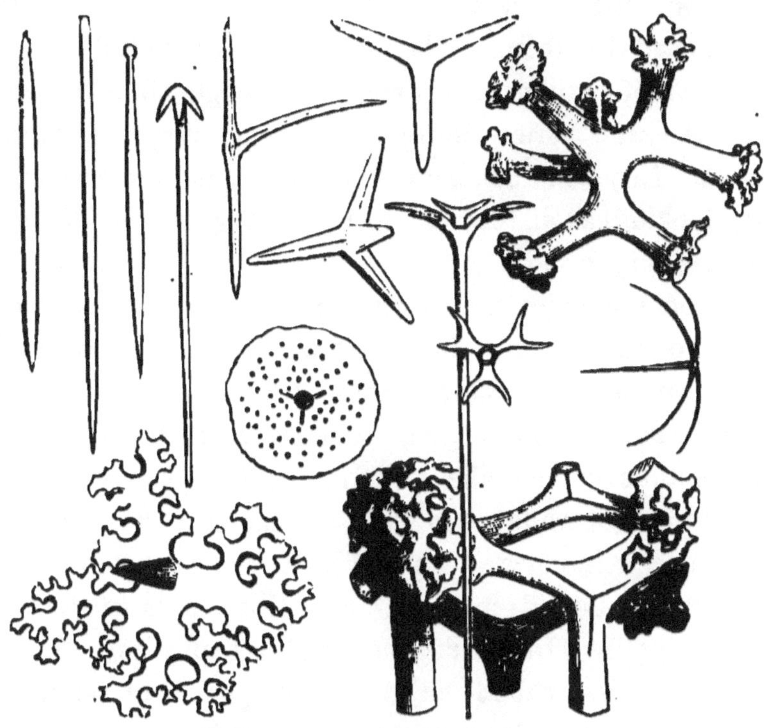

Fig. 30.—SPONGE SPICULES.

thus keeping up a constant circulation, by means of which all refuse and bad air are carried off; for though they cannot be said to breathe, they do take up oxygen, and give off carbonic acid.*

* Professor Huxley compares the sponge with a kind of subaqueous city, where the people are arranged about the streets and roads in such a manner that each can easily appropriate his food from the water as it passes along.

The fibres of the skeleton are, for the most part, of uniform size, but here and there one is thicker than the rest, and contains a thorn of flint, called a "spicule." These are rarely found in the sponges sold for toilet pur-poses, but the skeletons of other kinds are frequently strengthened by them. Though usually of flint they are sometimes of carbonate of lime, and though all of micro-scopic size, the variety in their shapes is simply endless. There are fairy fish-hooks, crochet-hooks, stars, and toasting-forks, looking as if they were made of glass of different colours.   (Fig. 30.)

In some sponges the skeleton is entirely composed of spicules of carbonate of lime; and in many both fibres and spicules are of silica.   One of these latter, called "Venus's Flower-basket" (Fig. 31, A), from its great beauty, is a tapering tube curved like a horn, which looks exactly as if woven out of white spun glass, so regular is the network. When alive it is covered with a film of greyish-brown jelly, and lives half-buried in mud, into which it is prevented from sinking by a fringe-like root of glassy spicules.

Another kind, called the glass-rope sponge, or sea-whip (Fig. 31, B), has long roots of transparent, colourless silica, like threads of glass, some as fine as hair, others as coarse as twine, which are twisted into a coil about a quarter of an inch thick, and some twenty or more inches long.   The roots are buried in mud, and the soft sponge grows above, but the whole was thought to look so "un-natural" that, as the first specimens were brought from Japan, they were believed to be artificial productions. Fine sea-whips have, however, been found off the coast of Portugal and elsewhere, so there is no longer any doubt of their genuineness.

It is, however, the smaller and chiefly invisible or-

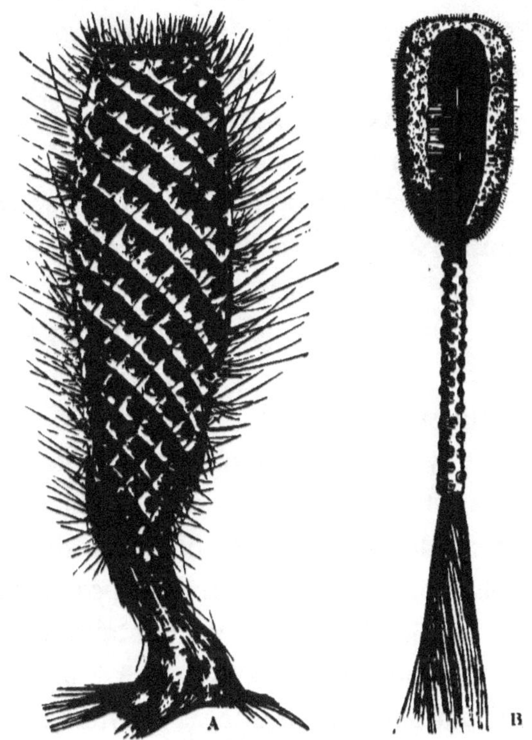

Fig. 31.—Sponges: A, Venus's Flower-Basket; B, Glass-Rope Sponge.

ganisms which use up most of the silica carried into the ocean, and which have left the most extensive remains.

Professor Bischof compares them with an extensive pumping apparatus, so enormous are the quantities of water perpetually passing through them. A hundred millions weigh less than one grain, but being full-grown in twenty-four hours, each individual must in that time imbibe at least 19·2 grains (about nineteen drops), or 33,333 times its own weight, of water, in order to obtain the silica required for its case or sheath.*   Nineteen drops !   The quantity sounds insignificant ; but if human beings drank at the same rate in proportion to their size, each would swallow five million pounds in twenty-four hours, or enough to drive a mill-wheel.

This microscopic population is so vast that we cannot form the faintest idea of its numbers, but we do know that one species is capable of multiplying four-fold in from twenty-four to thirty hours, so that in ten days a million individuals might spring from one parent.

But one little animal, the Rotifer, multiplies faster still, and in thirty days might have a trillion descendants, the weight of whose silicious sheaths would be 65,000 lbs. ; and supposing the mass to be of the density of " mountain-meal," it might form a bed of silica twenty-five square miles in extent, and about one foot and three-quarters thick.

Fig. 32.— Common Rotifer.

The Rotifers, or wheel-animalculæ (Fig. 32), so called from the wheel-like arrangement of *cilia* with which they are furnished, are much higher up in the scale of life than the Infusoria, though so minute that one species, having a single ruby-like eye, measures only $\frac{1}{125}$th of an inch in

* As already stated, 1,000 grains of sea-water contain but a minute quantity of dissolved silica, and as it is unlikely that the diatom or infusorian extracts the whole amount, they probably swallow much more than nineteen drops.

length and $\frac{1}{350}$th in breadth. They inhabit both fresh and salt water, and some have sheaths and some not.

The Infusoria are likewise furnished with *cilia*, by means of which they move and bring their food within reach, and having mouths, are superior to the Foraminifera and other Protozoa, or lowest forms of animal life, which can hardly be said to have any organs, and absorb their food through the whole surface of their bodies, which are but specks of slime or jelly.

Many of the Infusoria are covered by a sheath or shell, and most commonly increase by the simple process of splitting themselves in two.

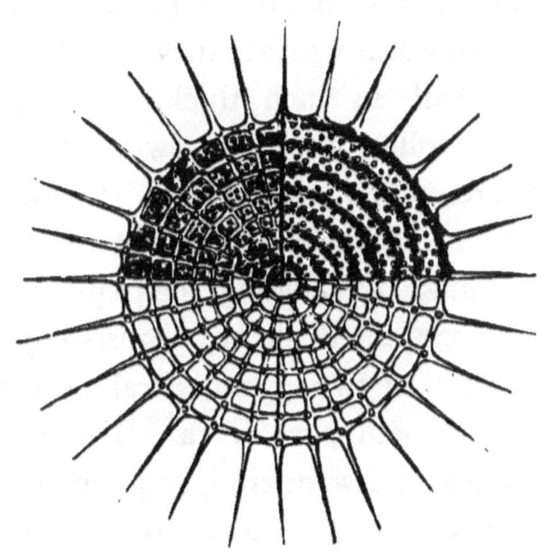

Fig. 33.—SKELETON OF RADIOLARIAN.

They multiply so rapidly that one (*Polygastria*) may in forty-eight hours have given rise to a progeny too large to be expressed in numbers.

Below the Infusoria, are the Polycystinæ, or Radiolaria, which are nearly allied to the Foraminifera, but have shells of flint often set all over with hair-like needles, the whole shell, needles and all, being a mere speck of the size of a grain of sand. (Fig. 33.) Some are extremely beautiful objects when magnified, and resemble the carved ivory balls for which the Chinese are noted. Notwithstanding their extreme minuteness, they have formed whole beds of silicious rock, such as that known as "Barbadoes earth," which is almost entirely composed of their shields.

Among the most important consumers of silica, however, are the microscopic plants called Diatomaceæ (Fig. 5), which swarm on the surface of the ocean, and are found, more or less, in all waters, salt or fresh, and in all latitudes.

In warm weather they sometimes form a mouldy covering, half an inch thick, on stagnant water; sometimes they collect in a yellow-brown layer at the bottom of ponds, or on water-plants, stones, mosses; or they may be found filling the towing-net in the Antarctic regions, and staining the floating ice with an ochreous tinge. They are one-celled plants, consisting usually, as their name implies, of two symmetrical portions, or valves, which are coated with pure silica. The variety in their shapes is endless and wonderful, and almost all are delicately marked and sculptured with bands of dots and lines.

Yet in spite of the exquisite beauty of their glass cases, they are so minute that 41,000,000,000 could be accommodated in one cubic inch of space, and 186,000,000 would weigh but a grain and look to the naked eye like a mere pinch of dust. They multiply rapidly, however, and one may become 8,000,000 in forty-eight hours, while in four days its progeny will fill two cubic feet of space.

It is these diatoms which are the main cause of the silting up of harbours. At Wismar, on the Baltic, they accumulate at the rate of 17,496 cubic feet annually, enough to form a solid cube measuring twenty-six feet each way. As they die, their cases fall through the water, in a constant, steady shower, and in the course of ages they have formed deposits many miles in extent and many fathoms thick. The thickest known deposit is that upon which Berlin is built, which is eighty-four feet deep; there is another eighteen feet thick beneath Richmond (Virginia); and one at St. Petersburg is thirty feet thick.

The "polishing slate" of Bilin, "mountain-meal" of Sweden and Tuscany, "Richmond earth," *kieselguhr*, or "flint froth" of Germany, the "Tripoli stone" of Africa, Italy, Bohemia, Germany, France, and the United States, as well as the beds of white earth on the banks of the Amazons, and the "Bath-brick" from the bed of the river Parret, in Somersetshire, all consist chiefly or wholly of the indestructible cases of diatoms and other minute organisms, some deposited in fresh, others in salt water. *

The "infusorial earth" of Nova Scotia—so called, though the vegetable nature of the diatoms has long been established—looks much like chalk, and is a very light, white, friable earth, which, when examined by a bright light, shows an infinite number of glistening specks of pure silica, whose hardness and sharpness make them useful for polishing.

"Electro-silicon" or "magic brilliant," the white substance employed for cleaning jewellery, is also a diatom earth found in Nevada.

In Lapland a similar earth is mixed with bark and used as food, and in America the farmers have been trying it as

---

* Rottenstone has a different origin. Carbonated water has filtered through beds of silicious limestone, carrying away the lime and leaving a light porous residuum of flint.

manure to supply the lack of silica. Some of these earths, whether of animal or vegetable origin, have been mixed with clay and lime, and made into " floating bricks," one-sixth the weight of ordinary bricks, and fireproof; some are added to sealing-wax, paper, soap, indiarubber and modelling-clay, to give consistency, and, finally, the once harmless little vegetables are mixed with nitroglycerine and converted into dynamite !

Nature acts more kindly by them, and is believed to have turned some of them into semi-opals.

One of the most important ends served by the diatoms, however, is that of being food for many of the lower animals, especially the Protozoa, among which we may mention the beautiful little Noctiluca, or "night-light," 30,000 of which may be contained in one cubic inch of water. They are themselves shell-less, moulded in the shape of melons, the largest hardly bigger than the head of a minikin pin, and are brilliantly phosphorescent. (Fig. 34.) At certain

Fig. 34.—Noctiluca Miliaris.

seasons they crowd the waves in such multitudes that one will be found in each drop of water. Phosphorescence in the open sea is, however, said to be produced chiefly by a minute plant (*Pyrocistis*) of the [size of a pin's head, which is very abundant far from land and, like the diatom, also has a thin casing of silica.

But diatoms contribute indirectly to the nourishment even of the great North Atlantic whales, for these feed chiefly on the acorn-like medusæ, brought to them by the Gulf Stream, which sometimes cover the ocean so thickly as to make it look like a prairie strewn with yellow leaves. Each medusa consists of from five to nine lobes, and as each lobe has been found filled with diatoms to the number of 700,000, each individual must swallow from three and a half to more than six millions at a meal.

Diatoms have another important function, which will be referred to later, but the countless millions which are for ever dying and sinking must form a considerable item in the bill of fare of many of the creatures which live at the bottom of the sea, and are therefore wholly dependent upon what is brought to them. For there are no beds of weed to which they can go for a meal. Seaweeds require a certain amount of light, and usually form only a fringe about a mile wide round the coast, being practically limited to depths of less than a hundred fathoms, though stragglers are met with here and there.*

The average depth of the ocean between 60° N. and 60° S. is 2,500 fathoms, and the population would be scanty indeed if they lived only on such pieces of weed as are torn from the coast. Besides the deep-sea dwellers, there are not only the Foraminifera already mentioned, but, in the hotter seas, vast multitudes of the wing-footed mollusks, called *pteropoda* from the fin-like lobes projecting from their sides, which live in and on the surface of

* The most highly sensitive photograph plates remain unaltered 166 feet below the surface.

the water far away from the seaweeds on the coast. (Fig. 35.)

In addition to the diatoms, however, the inhabitants of the ocean may feed on the great masses of floating weed, such as the Sargasso, or Gulf weed, which occurs in some parts of the Atlantic over a space of 3,000,000 square miles, in olive or golden patches of various sizes, from a few feet to several acres, with lanes of dark blue water between. There is a similar, but smaller mass of weed in the North Pacific; and such multitudes of fishes, mollusks, crustaceans, zoophytes, &c., dwell in and feed on these ocean prairies, that the destruction of one of them would probably occasion greater loss of life than the destruction of a large forest on land. The Giant Kelp, which grows at Tierra del Fuego to the length of 360 feet, is also sometimes met with in a floating condition.*

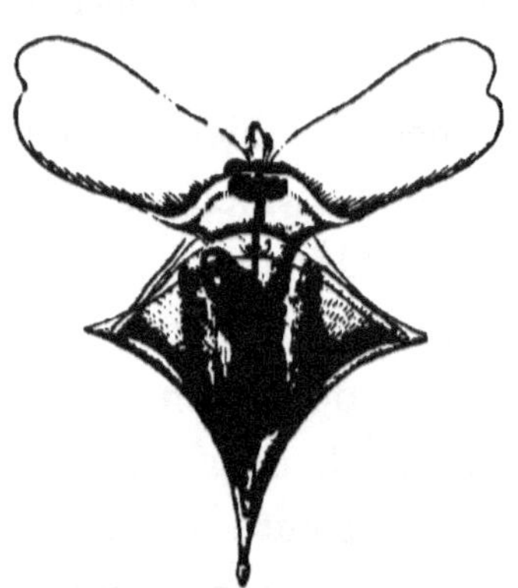

Fig. 35.—Pteropod or Wing-footed Mollusk.

Seaweeds differ from land plants in this, that they have no true roots, and the tough, leathery fibre or disk which takes the place of these, only clasps and does not penetrate the rocks, from which, therefore, it derives no nourishment. All their food comes from the sea, and much is supplied to them by the rivers.

* Lowly algæ also abound, and, indeed, the surface water of the open ocean is full of vegetable life.

Many seaweeds grow with great rapidity, and rocks which have been not only cleared but chiselled smooth, are found covered in less than six months with a dense growth of weed from two to six feet long, which must have sprung from seed. Besides silica and lime, they take up other minerals. Some take large quantities of magnesia; iodine, in combination with sodium and magnesium, is obtained from the ash of certain species, and some of the olive-coloured fuci contain so much sulphuric acid that where they are thrown up in large heaps and left to decompose, as is the case at Copenhagen, the silver in the neighbouring houses quickly turns black. Most sea-weed ash contains more or less soda, and that of the bladder fucus and some other olive seaweeds, yields so much that in former times many people made their living by it. The weeds were burnt in pits, and the dark-brown, half glassy-looking cakes of ashes, called "kelp," were sold for glass- and soap-making, bleaching, and other purposes. Scotland and her islands at one time supplied 20,000 tons of kelp annually; but the kelp-burners do little now, as the carbonate of soda, or soda ash, is now manufactured in enormous quantities directly from sea-salt. How does the salt, which sea-water contains in such abundance, get into the sea? Like the other minerals, it is brought by the rivers; but this some people find it difficult to believe, since river-water is fresh, not salt, and they prefer therefore to adopt the old notion that there are vast beds of salt somewhere in the ocean. These beds have not been discovered, however, and are therefore, at present at all events, wholly imaginary, whereas it is no imagination, but an ascertained fact, that all spring water, and almost all fresh water, contains a small quantity of salt.

Common rock-salt is a compound of the gas chlorine with the metal sodium, hence called chloride of sodium. Various compounds of chlorine (chlorides) are found in granite, mica, and many other minerals, as well as in volcanic rocks; and compounds of sodium exist in such enormous quantities that not a single minute speck of dust is said to be free from them.

Lava is frequently found covered with crystals of salt a few days after it has been poured forth; sometimes large masses of salt are thrown out, and after some eruptions of Mount Hecla enough has been found to load several horses.

There is no mystery, therefore, as to whence the rivers might be supplied with salt; but as the quantity they convey is usually too small to destroy their sweetness, it may at first sight seem strange that it should be enough to make the sea so intensely salt. But if only a small quantity of salt is being constantly added, while a much smaller quantity is withdrawn, any body of water must become perceptibly salt in time. We know that river water contains a minute quantity of salt, though our taste is unable to detect it. Sometimes, however, we can even taste it.

During the great drought in La Plata, which lasted from 1827 to 1830, when the land became so loaded with dust that old landmarks were obliterated, and much confusion and even lawsuits were the consequence, evaporation went on rapidly, and as, in addition to this, the rivers were fed entirely by springs, many of the small ones became so salt as to kill the animals which drank of them.

Evaporated water, as has been said, is almost pure ; chlorides and other * salts and minerals are left behind and accumulate, so that the smaller the body of water became the salter it would be. This is what has taken place in the Great Salt Lake of Utah, which was formerly much larger and filled with fresh water, as is evident from the remains of fresh-water shells on its ancient beach. At that time it had an outlet into the Snake River and so to the Pacific, and its waters were kept in a constant state of circulation ; whereas, now that it has shrunk and has no outlet, all that is brought into it must needs remain, with the exception of the water, which is constantly evaporating, and that, too, a little faster than it is poured in by the feeding streams. Consequently, the lake must still be growing smaller and salter.

The water is in fact far salter than that of the sea, being saturated, *i.e.*, it contains as much as it can hold, and deposits of salt are formed on the cliffs and rocks. If ever the whole body of water should be evaporated, beds of pure rock-salt would remain behind.

Salt beds are forming at the present day on the borders of the Black Sea, where all the rivers between the Danube and Dnieper widen out into shallow lakes, which are separated from the sea by narrow dams. Some of these lakes become partially dry every summer, and salt is then deposited in thin layers round the margin, but often as much as a

---

* A salt is the substance formed when an acid combines with an alkali and forms a body neither acid nor alkaline, but neutral. Thus :—

  Sodium and sulphuric acid make glauber salts.
  Sodium and nitric acid   ,,   Chili saltpetre.
  Potassium and nitric acid   ,,   nitre or saltpetre.
  Calcium and sulphuric acid ,.   gypsum or calcium sulphate.

L

foot thick in the middle. As much as 216,000,000 pounds were obtained one year, for one of the lakes is forty miles wide.

Supposing communication were cut off between the Black Sea, Mediterranean, and Atlantic, then as the water removed by evaporation would exceed that received by the rivers, the seas would become salter and salter until the water was saturated, after which all the salt which could not be held in solution would simply fall to the bottom and there accumulate. Something of this sort has probably taken place in the Dead Sea, which has nowhere any visible outlet.*

The salt we use with our food is manufactured from the Cheshire and Worcestershire brine-springs, which have flowed through beds of salt and are in many cases almost saturated. They are known to have flowed for 1,000 years, and as 1,630,000 tons of salt have been obtained in one year from the Cheshire springs alone, they must have conveyed enormous quantities to the Mersey, and so back to the sea. Much of the Cheshire rock-salt is so pure as to need nothing but crushing before it is fit for use.

Another mineral often found with rock-salt is gypsum, or sulphate of lime, which is the first to separate and fall to the bottom when sea-water is evaporated. Beds of salt and gypsum two feet thick have been formed on the rocks in some places simply by the evaporation of spray.

Gypsum, when burnt, and thus freed from water, is called plaster of Paris; this is made into a paste with water and used for making casts and moulds.

If all the invisible salts of the sea could be extracted, they would cover several millions of square miles one mile

* Sea-water is not nearly saturated.

deep ; and yet all this vast amount of matter makes no difference in the bulk of the ocean, whose waters would occupy
as much space if it were all taken away. It does, however,
affect the weight of the water, which is much heavier than
fresh water, and is heaviest where it is saltest.

So perfectly is the circulation of the ocean maintained,
by means of currents, &c., that sea-water varies little in
composition all over the world ; and even before the cutting
of the Suez Canal, the waters of the Mediterranean and Red
Sea contained the same minerals, though the latter is not
fed by a single river or brook, and communicates with the
Indian Ocean, while the former opens into the Atlantic.

As, however, evaporation takes place from the surface,
the surface water becomes slightly salter, but as it also becomes heavier, it sinks, the lighter water from below taking
its place, and as a constant exchange is thus kept up, no one
part can be much salter than another.

In the tropics, where evaporation goes on rapidly, and
on coasts where much water is locked up in the form of ice,
which is always fresh, a slight increase of saltness is observ-
able.

CHAPTER XI.

VEGETABLE SCAVENGERS.

Drowned Fishes—Burning and Breathing—How Fire may be Extinguished —Carbonic Acid Gas: how produced, how disposed of; taken up by Leaves—Quantity required by Beech Forest—Oxygen returned to the Air; Seaweeds as Scavengers -Carbon in Vegetable Oils, Acids, Sugars, Starch, Gums, Perfumes—Other refuse Gases—Fungi as Scavengers—Fermentation and Putrefaction—Floating Matter of the Air.

WHEN Professor Nordenskjöld was in the Arctic Ocean he noticed that in one spot there were innumerable dead fishes. A shoal had evidently been entangled among the ice, and becoming enclosed in a narrow space, had been *drowned !*

"But how can fishes possibly be drowned?" some one may ask. Well, a man drowns for want of air, if his head be kept under water, for drowning is practically the same thing as suffocation ; and the fish, though it lives in water, needs air as well as other animals, and being unable to decompose the water, and so obtain oxygen, is dependent upon that which is dissolved in the water. When that is exhausted it must needs be suffocated.

But why should not the same air be breathed over and over again ?

On one occasion M. Huber, having closed the entrance of a bee-hive, noticed that, in a quarter of an hour, the bees became uneasy, ceased working, and began vibrating their wings in great agitation. In about thirty-seven minutes they were exhausted and fell down ; thousands strewed the floor of the hive, and all would have been suffocated had the experiment been continued. As it was, the admission of fresh air revived them.

Evidently, therefore, the breathed air had been so altered as to be incapable of supporting life. When air passes into the lungs of an animal, part of the oxygen unites with the carbon present in the blood and oxidises or burns it, with the result that carbonic acid gas (carbon dioxide) is formed.* A similar thing happens when we burn wax, tallow, oil, gas, &c., all of which consist of carbon and hydrogen variously modified. Part of the oxygen of the air unites with the carbon, producing carbonic acid, and part with the hydrogen producing water. The latter we can see by covering a bit of lighted candle with a tumbler, which will at once become clouded with dew.†

But the flame burns, as the animal breathes, only while it has enough oxygen, and as there is but little in a tumbler, it goes out in a few moments. Not that the oxygen is entirely exhausted, but the proportion being too small, the flame is stifled. Hence carbonic acid may be used for extinguishing fires, and one which had been burning in a coal mine thirty years was, in 1851, successfully smothered by this means.

This carbonic acid is the refuse we are now to consider. This gas is one and a half times as heavy as the air, so

* One atom of carbon and two of oxygen make a molecule of carbon dioxide.

† When a lamp is first lighted, the glass, being cold, is similarly clouded, until the dew is evaporated by the heat of the flame.

heavy that it can be poured from one vessel into another ; yet instead of resting upon the earth, as we might expect, it mounts up and mingles with the other gases of the atmosphere.

By heating twelve grains of carbon* in thirty-two grains of oxygen we produce forty-four grains of colourless gas, which when frozen resembles snow.

Ordinary air contains about $\frac{1}{2500}$ of the gas, but there is much more in the air of our rooms, owing to the presence of fires and lights, as well as our own breathing. The air we breathe out contains from three to six per cent. of carbonic acid, and not enough to keep a candle alight.

A man breathes out about a cubic foot † of the gas in an hour and a half, and as one in a thousand is enough to make the air unwholesome, his breath alone would make the air of a room, measuring ten feet each way, unfit for respiration in that time, supposing it to be air-tight. He would be able, indeed, to go on breathing some time longer, but would feel drowsy and heavy; two per cent. of the gas would give him severe headache (even one per cent. is not long to be endured), and ten per cent. would stop his breathing altogether.

Fortunately our houses are not air-tight, but if we close our doors and windows, cover all cracks and crevices with list and sandbags, shut down the register, or put a chimney-board before the fire-place, and then sleep in a room ten feet square, the amount of fresh air which can come in is so small that we must needs breathe what is unwholesome during the

---

* The "black-lead" of our pencils is pure carbon, so also is the diamond.

† Nearly four and one-third gallons per hour.

greater part of the night, and it is no wonder if we awake with a headache and feeling unrefreshed. The case is worse still if we keep the gas burning, or even "just a bead," as some are fond of doing.

Those who live in towns are so accustomed to breathe air which is not quite pure that they do not notice it unless it be worse than usual, but an Arab, fresh from the pure air of the desert, wears a frown of disgust when business obliges him to enter a town, and usually goes about with cotton in his nostrils or a handkerchief drawn over his nose, and if obliged to pass a night within the walls, will at least not sleep under a roof.

It is computed that Manchester, with its large population and numerous furnaces, engine-fires, &c., produces some 15,066 tons of carbonic acid gas every day ; and the inhabitants of London send up some 800 tons of carbon into the air, in the same time. Yet all this great mass is quite invisible and has no share in producing fogs or darkness, for the human furnace is happily so constructed as to consume its own smoke. Each full-grown person contributes on an average more than four ounces of solid carbon to the atmosphere in twelve hours.

But the gas is also generated by that slow burning of animal and vegetable substances which we call *decay*, and it is poured forth in very large quantities by volcanoes and from cracks in the earth in volcanic districts. There are more than a thousand carbonic acid springs in the Eifel and Lake of Laach district alone.

The various processes by which carbonic acid is produced go on chiefly in the northern hemisphere, since that contains the larger proportion of land ; and as the Old World is more densely populated than the New, more carbonic acid must be produced in the East than in the West, in the North than in the South.

Yet the air all over the world, in the plains and on the mountains, is very nearly alike, for gases possess the peculiar property of intermingling or " diffusing " themselves equally through one another without any reference to their comparative weights. In this they are very different from liquids, such, for instance, as oil and water ; for the oil will always rise to the top even if put in at the bottom of a bottle. But if you put in first carbonic acid, which is the heaviest gas, then oxygen, and finally hydrogen, which is the lightest of all, in a little while, without any shaking, they will have thoroughly mixed with one another.*

Carbonic acid is forty-four times as heavy as hydrogen, and 1·529 heavier than air, so that but for this law of " diffusion " it would rest upon the earth ; and if all the carbonic acid breathed out by four million Londoners were thus to settle down, they would soon be completely enveloped in it and suffocated. As it is, however, except in particular localities, the amount of carbonic acid present in the atmosphere is nearly everywhere the same, and though it forms but a small proportion of the whole, its entire weight amounts to more than three billion tons.

Though constantly receiving additions, the quantity does not increase, thanks to the innumerable scavengers always at work removing it. Each blade of grass in the meadow, every leaf in the field, wood, or forest is busy during every

* Hydrogen is 14½ times lighter than air.

moment of sunshine in drinking in this, to us, poisonous gas, and they do this so rapidly that there is actually rather less of the gas close to the earth's surface, where it is generated, than there is higher up.

One plant of colza-rape, for instance, will drink more than two quarts of the gas in the course of one day's sunshine. An acre of beech-forest takes about three and a half tons of gas, or one ton of solid carbon, every year; and if the whole earth were covered with beech-trees, the supply of gas would be altogether exhausted in about eight years, supposing there were no means of renewing it.

As, however, three-fourths of the globe are covered with water, and as the vegetation of the fourth quarter is less than a third of what it would be if covered with forest-trees, the present supply would last a hundred years.

It is the leaves, not the roots of plants which take up carbonic acid ; for though the soil contains much, plants can be grown and brought to perfection in water which is quite free from it, provided, of course, the other food they need be supplied to them ; and they will be found to contain quite as much carbon as those grown in earth. They must therefore take it from the air, and as fast as they remove it from their own immediate neighbourhood, its place is supplied by more, in obedience to that law of diffusion already mentioned ; so that we may imagine streams of the gas to be constantly flowing towards every leaf and blade.

It must not be supposed that the breath and fires of England necessarily feed only the English crops and trees, for then what would happen during the winter months, when we have comparatively few scavengers to purify the air for us?

When it is winter here, however, it is summer in the South, and, thanks to winds and currents, as well as to the movement of the gases among themselves, our breath may go to feed the palms and sugar-canes of the tropics, while English oaks and English wheat are in their turn fed by the breath of Africans and South Americans.

When the leaves have attracted and absorbed the gas, the sunbeams with their rapid vibrations tear the atoms of carbon and oxygen asunder; the carbon is kept by the chlorophyll or leaf-green, to which the leaves owe their colour, and nearly the whole of the oxygen is given back, to be breathed over again and to purify the air of those organic impurities, which, besides carbonic acid, are constantly being poured into it from the lungs of animals.

Plants breathe by means of their leaves, and if stripped of them will die: but without the leaf-green they cannot separate the carbon, and to make leaf-green they need iron, in very small quantities, it is true, but if kept quite without it, the plant, like the human being, grows pale and sickly, it cannot digest its food, and finally dies of starvation. But the leaf-green, however healthy, cannot do its work without sunlight, and plants kept in the shade, or even in rooms lighted only from one side, give out carbonic acid instead of oxygen.

At all hours of the day and night plants give off carbonic acid, but the quantity is so minute that in the sunlight it almost escapes notice, whereas by night, or in the shade, they give off carbonic acid only. Even then, however, the quantity is so small that, as has been said, "one might safely pass the night in a greenhouse without any danger of being suffocated by the geraniums." Such plants as the colza, pea, bean, raspberry, and sunflower, exhale during a whole night only as much as they absorb during one quarter of an hour or twenty minutes of direct sunshine.

Seaweeds, as well as land plants, perform the office of scavengers, not only by absorbing carbonic acid, but by giving out oxygen; and Dr. Hooker, remarking upon the universality of the diatoms, speaks of them as a most important feature of the Polar seas, where the higher forms of vegetation are so scarce that the office of purifying the waters devolves mainly upon them.

Many fresh-water plants, among which the common duckweed is prominent, are powerful purifiers, whether their leaves float below or upon the surface; and those who have kept aquariums will, no doubt, have noticed the little globules of oxygen which cover the weeds when the sun shines upon them. By the oxygen thus evolved the impurities in the water are literally burnt up.*

The carbon taken up by plants is in some unknown manner combined by them with oxygen and hydrogen to make cellulose, the colourless material of which all young vegetable fibres are composed. It is to be seen in the skeleton leaves, sometimes picked up in winter, with all

---

* Mere exposure to the air will purify water to a great extent. Ozone, or condensed oxygen, generated by electricity, is a yet more powerful consumer of all putrescible matter in the air. The air of impure places is universally characterised by a want of oxygen, and the differences, though

the fleshy part decayed away, and only the bare bones left; it is seen again in calico, linen, and paper, which, white as they are, have only to be held near enough to the fire to be scorched, and the carbon is revealed at once. Cotton-wool, again, is almost pure cellulose.

About one-half the substance of all wood-fibre is simply carbon ; but it is not all used up in these ways, and the immense variety of vegetable substances, composed of the three elements, carbon, hydrogen, and oxygen, is amazing.

First there are the vegetable-oils, palm, olive, rape, lin-seed, &c. All are hydro-carbons, as these compounds are called, and the acids which give them their characteristic flavours are hydro-carbons also.

From the same materials, combined in different pro-portions, the sugar-cane, beet-root, maple, and mallow, pro-duce their cane-sugar, the vine and other fruit trees their grape-sugar, and the acacias their gum-arabic. From the same, again, the lemon produces its citric acid, the vine its tartaric, the sorrel and lichens their oxalic, the rhubarb

minute, are exceedingly important to health. According to Dr. Hartwig, the percentage of oxygen in the air of the

Metropolitan Railway (Underground) is	... 20· 70
East End of London	... 20·857
Pit of a theatre ...	... 20· 74
Hyde Park	... 21· 00
Coast of Scotland	... 20·999
Suburb of Manchester, wet day	... ·20·980
Ditto ditto dry day ...	... 20·947
Sitting-room	... 20· 89
Mine	... 20· 14

its malic, and the nettle the formic acid which makes its sting so painful.

All the vegetables used as food contain from forty to fifty per cent. of carbon, and foremost among these are the cereals, the starch, sugar, gum, and oil of whose grain are all hydro-carbons. One pound of flour contains on an average seven ounces of carbon.

The various pines and firs combine carbon with hydrogen only, and produce turpentine; the lemon, bergamot, pear, lavender, pepper, camomile, clove, &c., do the same, and even in the same proportions, and produce the essential oils which help to give them their special scents and flavours; while the laurel of China and Japan adds an atom of oxygen, and produces the white crystalline gum known as camphor.

Various tropical trees combine carbon with hydrogen, and form such juices as indiarubber and guttapercha, which, though white as they flow from the stem, turn black and solid with exposure to the air; and then other plants, again, such as the rose, Tonquin bean, meadow-sweet, and many others, convert the same two elements into the sweetest perfumes by the addition of oxygen.

But there is another gas of which the vegetable world relieves us, and makes good use. This is ammonia, which, like carbonic acid, is produced by the decay of animal and vegetable substances, and is easily known by its strong, pungent smell. It is a compound of hydrogen and nitrogen, and is always present in the air, but does not on an average amount to more than one part in fifty millions; and though, as it streams up from all decomposing organic matter, it is naturally more abundant in the air of towns and cities, the supply from this source falls far short of the demand. For all plants require nitrogen (and without it those used for food would lose their nutritive qualities); yet, though there is an unlimited supply of it in the air, they are unable to take it up with their leaves, except in the compound forms of nitric acid (nitrogen, hydrogen, and oxygen) and ammonia (nitrogen and hydrogen).

From two to twenty-one pounds to the acre are washed down from the air on an average every year; but a single acre of clover-hay requires about 108 pounds, and in twenty-eight bushels of wheat there are $45\frac{1}{2}$ pounds of nitrogen. The remaining quantity is, therefore, derived from the decay of animal or vegetable matter naturally present in or added to the soil.*

Vegetables, then, derive all their nourishment from the air and from the soil, and convert gases and minerals into food for the animal creation, which is altogether dependent upon them for its means of subsistence.

But there is one group of vegetables which are like animals in this respect, that they must have animal or vegetable matter to feed on. These are the fungi, and wherever they are seen there we may be sure is some decomposing organic matter.

Fungi do not contribute to the purification of the air by taking up carbonic acid, for they resemble animals in their way of breathing, as well as of feeding; but they are scavengers, and some of them, such as the mushroom, are very perfect ones, since they convert refuse into wholesome

* Chapter VII.

food.   Those which grow upon decayed wood are, however, very unwholesome, and others are not only poisonous, at least to man, but so offensive that they cannot be considered any improvement upon the refuse to which they have given a new form.   Mr. Cooke mentions one, the scent of which became in an hour or two worse than that of any dissecting room, and was perfectly intolerable until wrapped in twelve folds of thick brown paper.

Still, what is disgusting to us is doubtless savoury to others, for many of the beetle tribe are entirely dependent upon the fungi for food.

Wherever decaying vegetable matter of any kind is met with, there fungi are sure to be present, hastening on the process and growing so rapidly that a crop of toadstools will spring up and a puff-ball grow prodigiously in a single night.

They are not even particular about growing in the light, and are found in mines growing on the wooden pegs driven into the walls for purposes of measurement, and at one time there was a specimen on the woodwork of the tunnel near Doncaster which measured fifteen feet across.

No substance, animal or vegetable, comes amiss to them, and they will grow on and consume the defunct members of their own race.

In New Zealand there is a certain caterpillar which, after swallowing the spores of a fungus, buries itself in the ground and dies.   The spores take root in it and the fungus growing and absorbing the entire contents of the skin, takes the exact form of the creature, but always throws out a joint at the back of the head.   It looks exactly like a caterpillar with a twig growing out of it, and is as hard as wood.

The spores of the fungi, which answer to the seeds of other plants, are microscopic, even the largest of them, and the smallest are hardly visible when magnified 360 times. Their numbers are so enormous that one single plant may produce multitudes such as the mind cannot realise, and being so extremely minute, clouds of them are constantly suspended in the air, ready to settle and take root on any suitable soil, such as jam, paste, cheese, or even an old boot if left in a damp place.

As long as this blue-green "mould" grows on the surface of a mass of paste, it can obtain from the air the necessary oxygen; but if buried in it it does not die, for then it decomposes the starch of the flour, takes its oxygen, and gives off bubbles of carbonic acid. The "bubbling" which we call fermentation, is, according to Professor Tyndall, just "life without air."

The yeast-plant, another of these ferment-producing fungi, and the only one which can be said to be "cultivated," produces hardly any fermentation if allowed to grow on the surface, where it gets oxygen from the air, but if buried in the wort, it has to decompose the sugar in order to obtain it, and fermentation proceeds so rapidly that streams of carbonic acid flow over the sides of the vat. A drop of yeast, the size of a pin's head, increases enough to ferment a pint of liquid, alcohol, as well as the carbonic acid, and a small quantity of glycerine, being formed in the process.

Yeast mixed with dough has a similar effect. Part of the starch is transformed first into sugar, then into carbonic gas and water, and the gas in its efforts to escape makes the minute bubbles which render the bread light and spongy.

In making aërated bread, carbonated water is used instead of yeast, with the same effect, and with the advantage that there is no loss of starch, and no products of decomposition are left in the bread.*

There are many other vegetable-ferments, most of which are, however, unpleasant; some produce "maladies" in beer, some acidity, some putrefaction, according to their various ways of feeding; but though each has a flavour peculiar to itself, the character of all is essentially the same. When we like their effect, we call it "fermentation," when we do not like it, we call it "putrefaction."

The name of bacteria (Fig. 8) is given to a large variety of these organisms, which, though much lower down in the scale of life than the mildew on decayed wood, are yet nearly allied to the fungi. They are the agents of all putrefaction, and though no microscope is powerful enough to detect their germs, they abound in all the moist places of the earth, and being so inconceivably minute that a grain of pollen is gigantic in comparison, are not only easily lifted into the air by the wind, on a dry day, but are even drawn up with the water as it evaporates.

It is evident that all putrefaction is caused by them; for milk, meat, &c., which remain perfectly sweet for an indefi-

---

* The use of the word "leaven" to denote a *good* influence seems strangely inappropriate in the face of these facts and the circumstance that " leaven in the inspired writings is always taken as the type of naughtiness and sin " The most prominent idea connected with leaven is that of *corruption*—but this idea was not peculiar to the Jews. The priest of Jupiter was forbidden to touch flour mixed with leaven, and the Romans sometimes used the word " fermentation " for corruption. "The radical force of the word *matztóth*— unleavened—is sweetness or purity."—*Smith's Dictionary.*

nite time if the air be excluded or filtered through cotton wool, will be found swarming with bacteria in a few days, if exposed to ordinary air.

There are billions of them in the air of most London rooms and on all exposed surfaces, even on the money which passes from hand to hand.

When developed they are for the most part colourless and transparent, are constantly dividing and sub-dividing, and exist either singly or joined together in chains. Though in size they vary from the 350th to the 1,000th part of a *millimètre* (one *millimètre* is less than $\frac{1}{25}$th of an inch) they are exceedingly tenacious of life, like other of the lower organisms, and in the germ state can bear great extremes of temperature without being killed.

They may be boiled and they may be frozen, and though unable to germinate under these conditions, the life in them will not be extinguished, but, after lying dormant, perhaps for months, they will become active as soon as a favourable opportunity offers. Oil of hops and carbolic acid, however, they cannot withstand.

Many of the diseases which attack human beings and animals—splenic, typhoid, scarlet fevers, &c.,—have now been clearly traced to the agency of bacteria, and it seems probable that whooping-cough,* measles, chicken-pox, and other infectious disorders, are also the result of their attacks. They increase with extraordinary rapidity, one germ being capable of producing, in twenty-four hours, 16,000,000 bacteria, which, when fully developed, can only just be seen

---

* The fumes from carbolic acid, sprinkled on a heated shovel, are found to cure whooping-cough very quickly.

by the aid of the most powerful microscope and with all the appliances of artificial light.*

It is no wonder, says Dr. Fürst, that infection should spread, when we realise that the air of a sick-room is loaded with germs so minute that hundreds may adhere to the smallest particle of skin, that they may settle in the hair, on clothes and books, may be carried away in numbers by the flies and deposited on food or anything else upon which they next alight, and, unless expelled by carbolic acid, may remain for months in carpets, bedding, &c. Even a bright silver spoon which has been used by a sick person will probably have germs adhering to it until properly disinfected.

People are more alive to the dangers of infection than they were, and Dr. Fürst mentions, as an instance of official caution, that a *telegram* sent from Alexandria during the last cholera epidemic, was detained twenty-four hours on the way, to be properly fumigated !

But we must not regard even the bacteria as purely useless or mischievous. "They are," says Dr. Tyndall, " noxious, like many other things, only when out of their proper place. In their place they exercise valuable and useful functions as the burners and consumers of dead matter, animal and vegetable. They are not all alike, and it is restricted classes only that are dangerous to man. There is no respite to our contact with the floating matter of the air, and the wonder is, not that we should occasionally suffer from its presence, but that so small a portion, and that diffused over wide areas, should be deadly to man."

* Thirty thousand million cholera-bacteria, or comma-bacilli, as they are called from their shape, occupy the space of a pin's head.

CHAPTER XII.

VEGETABLE REFUSE.

Petrified Wood—Rapid Decay of Vegetable Matter in the Tropics—Formation of Peat—Dismal Swamp—How Coal was Formed: Jungles, Luxuriant Vegetation, Drift-wood—The Club Moss, its Sacs and Spores—"White Coal," Brown Coal—The Pitch Lake; Bituminous Shales, Rock-oil—Diamonds, "Black-lead," Amber, Iron Pyrites.

VARIOUS are the fates of the vegetable scavengers, when, their time of active service in that capacity being over, they themselves fall under the head of "refuse," and pass into the lumber room to be remodelled.

Both at the Cape and in Australia, bushes near the shore are often killed by the calcareous sand which buries them. As they decay, some of their carbon is converted into carbonic acid, which dissolves the lime immediately around them, and as evaporation proceeds, this is re-deposited and forms a solid crust round the bark. When the decay is complete, and all the wood converted into gases and ash, only a pipe of limestone remains, surrounded by loose sand; and often these pipes are filled with hard calcareous matter, which makes them so solid that when the sand shifts and they are left exposed they look, especially when branched, like the white, stony skeletons of the shrubs they represent.

Sometimes vegetable substances are *petrified*, that is, as they decay and the various atoms of which they are composed are set free and returned to the air, the place of each one is taken by an atom of something else, and if the process is allowed to go on, the whole may be converted into, or rather replaced by, a mineral or metal, which forms a model so exact that all the minute fibres and cells and even the very texture of the original substance are clearly distinguishable, though none of it remains.

Pieces of petrified wood, in which the grain is distinctly visible, are found in large quantities in the Suffolk Crag, but petrifaction or mineralisation is not the common fate of dead vegetable matter. When left freely exposed to the air, it is slowly oxidised or burnt up, the hydrocarbons, cellulose, and starch of which, whatever its nature, it chiefly consists, being converted into carbonic acid and water.

The light-coloured fibres of the stems and leaves are gradually converted into a brown or black powdery substance, whose weight and bulk continually diminish as more and more carbonic acid and water are produced ; and, with a suitable temperature and full supply of air, the process goes on without interruption, though more and more slowly, until nearly all the carbon and hydrogen have returned to the air whence they came. The nitrogen, too, has been set free and gone back, partly as nitrogen, partly in combination with hydrogen as ammonia.

In the island of Trinidad, where the heat is perpetual and rainfall large, all vegetable fibre decays so rapidly that, as Mr. Kingsley has said, there is hardly a dead stick or leaf to be seen even in the primæval forest. An English wood, if left to itself, would be cumbered with fallen trees, and in North and South America there are forests, in the temperate zones, which are piled ten or fifteen feet high

with dead or dying trunks, in every stage of decomposition.

But Trinidad is little more than 10° from the equator,
and in that fierce heat fallen timber melts away in a few
months, or even days, and its gases, being rapidly absorbed
by the luxuriant vegetation around, enter at once upon a fresh
career.

In temperate regions, on the other hand, the fallen leaves
accumulate year after year, and so quickly that, although
the first stage is rapidly passed through, and leaves, twigs,
and sticks, are soon partially decayed, a fresh fall speedily
follows, and as this to a certain extent excludes the air, the
process is checked. It does not cease, but it proceeds more
slowly ; and whereas in Trinidad the soil, wherever visible,
is just a yellow loam, undarkened by leaf-mould, in an
English wood there is often a foot or two of elastic brown,
peaty soil, and in the Himalayan forests this leaf- and
timber-mould often accumulates to the depth of fifteen or
twenty feet.

In the Falkland Isles, where the climate is damp and
cool, almost every kind of plant, even coarse grass, is converted into peat, which is sometimes twelve feet thick, and
so compact that it will hardly burn.

It is in the cypress swamps of America, however, that
the formation of peat proceeds on the largest scale.
The Great Dismal Swamp of Virginia, which extends over
1,000 square miles, is covered with many kinds of shrubs and
trees—such as the white cedar or "cypress"— which like a
watery situation, as well as with water-plants without number,
and a multitude of ferns, reeds, &c., of all sizes up to

eighteen feet, while the surface of the huge quagmire is in many places covered by a layer of moss four or five inches thick. The black, spongy soil beneath is made by the decay of countless generations of these various forms of vegetable life, but it has advanced beyond what is strictly speaking the peat stage, for almost all trace of fibre has disappeared and the whole has been converted into mud, which again is dissolved and gives its colour to the clear, brown-tinted water of the pools.

The whole delta of the Mississippi (14,000 square miles) is for the most part covered by a series of similar swamps and similar vegetation growing with the utmost luxuriance, both in the water and in the black soil. In 1812 a large extent of "cypress swamp" in the Mississippi valley was disturbed by earthquake, and sank below the level of the water, peat, ferns, tree-stumps, fallen trunks, and standing trunks being covered with the river-mud or alluvium, which is brought down in such abundance. In the course of time the mud would so accumulate as to rise to the surface, and would speedily be covered with a fresh growth; another mass of peaty mud would accumulate on the top of the old one with a layer of hardened mud or sand between. This alluvium being, however, more or less porous, would not prevent the buried peat from parting with more and more of its gases; but as the hydrogen, oxygen, and nitrogen, would pass away more rapidly than the carbon, the proportion of this latter would continually increase, until there might be perhaps as much as eighty-two per cent. of carbon, with five and a half of hydrogen, and scarcely twelve and a half of oxygen and nitrogen together.

The mineral composition of the mass would, in fact, be identical with that of coal; and much coal has evidently been formed in the manner just sketched, for we can see the soil in which the ancient trees and plants grew, as well as the soil heaped on their heads; and the trees themselves

Fig. 36.—DIAGRAM OF COAL-SEAM AS SEEN IN THE FACE OF THE WORKING OF A COAL-MINE.

(*a*) Under clay with roots passing through it; (*b*) bed of coal; (*c*) "roof" of the coal, composed of sand and shales, with upright trunks of trees passing through it.

are often found standing upright and still rooted in the spot where once they flourished (Fig. 36).

Coal is found in the Manchester coal-field at intervals through a thickness of 6,800 feet, 60 feet of which are workable coal; but the seams are separated by so many other deposits that the swamp, if swamp it was, must have undergone many such subsidences as that just described.

Coal-seams, however, appear more often to be the remains of dense jungles growing along the coast, such, perhaps, as the mangrove-swamps of the present day. In the tropics great quantities of trees and shrubs grow down not only to the very edge of the salt water, but actually in it. Many of the coal-bearing strata accordingly contain marine shells, and there are in Russia coal-beds which alternate with limestone, showing that they must have been buried in the sea.

During the Coal Period, when the best coal and all our English coal was formed, the climate greatly differed from what it is now, and tropical plants flourished all over the world. Vegetation was far more luxuriant and its character was very different, for the English jungles were filled chiefly with ferns, reeds, horsetails, club-mosses, of gigantic size, but with no solid trunks ; and, indeed, there was but little wood, properly so called, in these old forests.

Another change, too, has taken place since man came upon the scene, for the river-banks are now covered with grass, corn, &c., which are carefully harvested, whereas, when the greater part of the land was clothed with a rank vegetation which was left to grow and decay as it would, every little brook and stream would be loaded with dead leaves, and in time of flood would tear up and carry away many a tree. The state of things must, indeed, have been somewhat similar to that now existing in the vast un-broken forests through which flow the mighty Amazons and its tributaries.

The Rio Negro and many other large tributaries are quite brown or even black, like bog-water, from the dissolved vegetable matter which they contain, while the little brooks are half choked with dead leaves, rotten branches, &c., and when, during time of flood, the great river rises forty or fifty feet, the largest forest-trees are uprooted and hurried away.

Enormous quantities of drift-wood are also carried into the Gulf of Mexico, some of the trunks having travelled 1,000 miles down the Mississippi. No doubt also large matted masses of plants, tree-ferns, &c., were carried down by the rivers or washed away from the low jungle edge of tropical islands, and either accumulated in estuaries or were thrown up on the swampy shore or carried farther out to sea. But wherever they accumulated—at the bottom of sea, lake, or swamp—there, if the conditions were favourable, they would be converted into coal.

The same may be said of seaweeds; for though no beds, except some small ones in Iceland, can actually be proved to be formed of them, yet seaweeds, as well as shells, are found fossil in coal, and there is every reason to believe that they have contributed to its formation, for we know that the ocean teemed with inhabitants which must have been fed as at the present day.

There is one plant, however, which has contributed so largely to the making of English coal that a few words must be devoted to it. This is the common ground pine, or club-moss, now an insignificant little plant only a few inches, or at most two or three feet, high, the stems of which are covered with little scale-like leaves and terminate in spikes resembling fir-cones. Between the leaves of the spikes are small round bags filled with fine dust, like the pollen on the anthers of flowers, which consists of spores, and is so resinous and inflammable that it has been used in the preparation of fireworks. A pinch of this lycopodium powder burns in a candle with an instant flash, and being not readily affected by water is used by chemists for coating pills.

In the Coal Age there flourished giant relations of the little club-moss, forest trees in fact, some of them a hundred feet high, which might easily have been mistaken for pine trees. Their cones, or rather catkins, did not, however, produce seeds, but the leaves composing them bore on their surfaces little sacs, or cases, scarcely larger than those of the present club-moss, and, like them, filled with spores.

A thin slice of coal under the microscope is seen to contain multitudes of yellowish-brown bodies, about the twentieth part of an inch in diameter, which are flattened bags, often filled with irregularly rounded hollow bodies, measuring about one-seven-hundredth part of an inch across. These are the spores and their cases, which form by far the larger part of all the English bituminous coal examined by Professor Huxley, and contribute mainly to its inflammable character.

Clouds of yellow dust may be shaken from a branch of club-moss, and on the shores of some of the Canadian lakes there are often great heaps of yellow pollen which has been blown from the pine-forests, and being too light to sink, has been thrown up on the muddy shore by the waves. If this should ever be consolidated it would form, with the mud, a sort of shale very similar to an ancient deposit on the shore of Lake Huron, which is so full of spores and cases that it burns readily with a bright flame. (Fig. 37.)

There are innumerable seed-like bodies in the curious "white coal," as it is called, though really it is brown, now forming in Australia, which has all the character of true coal,

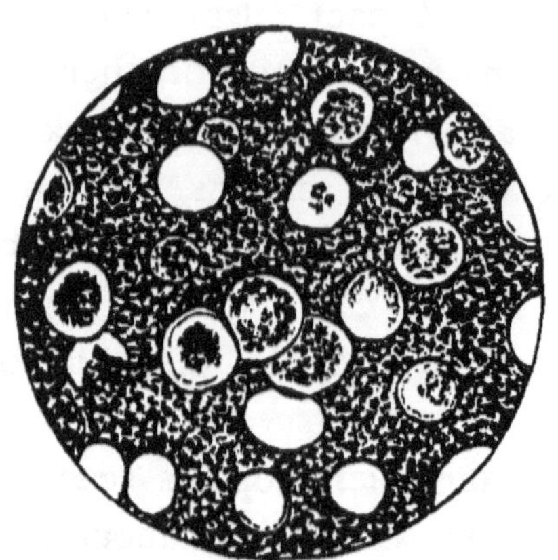

Fig. 37.—THIN SLICE OF SHALE FROM KETTLE POINT, LAKE HURON. (*Greatly magnified, and showing the little globular spore cases scattered through it.*)

though of inferior quality, owing to the clay and sand mixed with it.

The combustible portion consists entirely of spores, as is the case also with the "Tasmanite" of North Tasmania, which forms a stratum several feet thick and some miles in extent, and has a granular appearance, owing to the multitude of little round bodies it contains.

The "black shales" of Ohio, again, which are more ancient than the Coal Age, contain a considerable percentage of organic matter, made up entirely of spores and cases. The coal of Borneo, on the other hand, is a much younger deposit than any English coal, and of quite a different origin, being formed from a mass of huge timber, half of which still shows the grain of the wood.

Old timber left in a mine in the Hartz which is some 400 years old has been found converted into brown coal, or lignite, an earthy-looking, lustreless substance, of which large deposits are found in many parts of the world, especially in Hungary. Brown coal is far more ancient than this Hartz-mine timber, but is modern compared with true coal, and contains a smaller percentage of carbon, but as it continues to give off gases, there seems no reason why it should not turn into true coal at last.

Even then it will not cease to give off gases, two of which are often fatal to the miner, one being the deadly "choke-damp," or carbonic acid, which collects in old and ill-ventilated mines; and the other, the "fire-damp," a compound of carbon and hydrogen, which ignites with violent explosion on the introduction of a light and the admission of a certain proportion of air.

Coal varies extremely, according to the amount of gas it contains, and when at last little is left but carbon and ash, it is called anthracite, which is so hard as not to soil the fingers, and burns with a dull red, flameless, and smoke-less glow, giving out great heat.

In the Wallsend Colliery so much inflammable gas escapes from the coal, that on the insertion of a tin pipe in a hole drilled for the purpose, and surrounded with clay, the gas issuing from the pipe may be lighted just as at an ordinary gas jet.

It is by heating coal in large closed retorts that the gas we burn is obtained, and the same thing may be done on a small scale by filling the bowl of a tobacco pipe with coal-dust, covering it close with clay, and holding it in the fire. Enough gas will be driven up the stem of the pipe to burn when a light is applied.

Nature has in some cases used probably heat, and certainly pressure to drive off the gas from large masses of coal, for in the Appalachian coal-field, which covers at least 63,000 square miles, according to Prof. Bischof, the coal passes into anthracite where the strata have been most disturbed by earthquake; and one remarkable bed, now fifty feet thick, is computed to have been probably two or three hundred feet thick before it was thus compressed.

And now to consider some of the other metamorphoses undergone by dead vegetable matter.

Landing on the point of La Brea, in the island of Trinidad, we see that the beach is black with pitch, and that the so-called rocks, or reefs, consist also of pitch. One of the "rocks" indeed has been almost dug away to make asphalt pavement in New York and Paris. The "pebbles" on the shore are of the same substance, and everywhere there is pitch, no more than a foot or two beneath the surface, which is constantly oozing through the brown soil, itself half pitch, in which the pineapples for which La Brea is famous grow to especial perfection. A pitch road leads up to the famous Pitch Lake, a mile and a half in circumference, which glares and glitters in the sunlight, and has something the appearance of a bed of gigantic black mushrooms of all shapes and sizes, from ten to fifty feet across, their round heads pressed close together, and the spaces between filled with water. In one part of the lake the pitch is constantly oozing up from the depths below, and is quite liquid; and the ground everywhere in the vicinity is full of pitch and coal-like matter to the depth of hundreds of feet. Beneath the lake there is said to be a bed of coal, and it is this coaly matter, as it seems, which is constantly turning into pitch and oil, which are forced up through every crack by the enormous weight of shale and sandstone above.

The pitch is everywhere hardened into asphalt from the evaporation of the oil, except in the centre, and its highest temperature does not exceed 35° C. (95° Fahr.). There is no evidence of any volcanic eruption close to the lake; but there is an active mud-volcano twenty miles off, and the mainland is often shaken by severe earthquakes, so that in all likelihood the pitch is derived from beds of vegetable matter, which are being slowly distilled by volcanic heat. A mile or two off there are beds of brown coal, one of which, if continuous, would pass beneath the lake at a great depth; and when it is considered that for ages past the Orinoco has been rolling down vast quantities of timber and vegetable matter, it seems highly probable that these are the materials from which the pitch is made. It is certainly derived from decayed vegetable matter; and sticks which drop into the liquid pitch are often found partially transformed into the same substance.

Another product of the decomposition of vegetable matter is rock-oil or petroleum, which is found in many parts of the world. Bituminous shales—*i.e.*, hardened mud impregnated with vegetable matter—are also made to yield oil, and thus, thanks to the way in which we have learnt to utilise nature's refuse, estates formerly worth but a few hundred pounds a year, now bring in as many thousands. ·

There are about a hundred petroleum wells in Burmah, and there is one in Zante which has been flowing for 2,000 years; but petroleum was not discovered in America and Canada until 1861, in which year it was mentioned as an important fact that 250 barrels were exported. And important it undoubtedly was, for by 1882, the exports of petroleum from New York and Philadelphia had considerably exceeded 500,000,000 gallons a year.

It is not improbable, however, that the petroleum of North America may be in part of animal origin. American oil has been exported to all parts of the world, and until recently held almost exclusive possession of the European markets; but, now that the still vaster quantities about Baku, on the Caspian, and throughout the whole region north of the Caucasus are being explored and systematically worked, the American oil seems likely to be in a measure at least superseded.

But whether we burn coal, rock-oil, gas, or any of the various kinds of candles, our rooms are warmed and lighted by the heat and light drawn from the sun by trees, ferns, mosses, &c., ages ago. The carbon taken from the air is sent back in the form of carbonic acid, and in one way or another all that was drawn from the air is sent back to it to feed new generations.

There are a few other forms of vegetable refuse, however, which we must not quite pass over.

At one time it was believed that the diamond, the hardest of all known bodies, was simply a vegetable gum, for it consists of pure carbon with a trace of ash, and its value therefore arises, not from its composition, but from the form in which it crystallises. Very minute but real diamonds have been artificially obtained from a gas containing carbon and hydrogen, the carbon being induced to crystallise by a method which need not here be described These diamonds are therefore of vegetable origin, but we are entirely ignorant as to the way in which diamonds are formed in nature.

Carbon crystallises in yet another and totally different form, namely, in black, opaque, six-sided plates, when it is called graphite, plumbago, or "black-lead." This may very well be derived from the decay of vegetable matter; and jet is considered to be a highly bituminised wood, which, from

Fig. 38.—ANIMAL REMAINS IN AMBER.

the fact that it is often found surrounding fossils, &c., seems to have hardened from a plastic if not liquid condition. The rocks in which it is found are often strongly impregnated with petroleum.

Amber is the resin (Fig. 38) of some extinct species of pine, and is often found with coal or fossil wood. Many pines and firs at the present day have resin between their annual rings; and large masses of gum are found at the roots of the New Zealand Kauri pine, and exported to the amount of several thousand tons every year for the manufacture of varnish.

N

Similar lumps are found at the foot of the Brazilian copal tree.

It may not at first sight be obvious what connection iron pyrites has with vegetable refuse; but being a compound of iron and sulphur, both of which enter into the composition of plants, it may clearly be derived from them.

The ash of beech-wood, for instance, contains enough sulphuric acid and peroxide of iron to form pyrites to the amount of $\frac{1}{48077}$th of the weight of the wood; and twenty-three times as much as this might be made if during its decomposition it should come in contact with water containing sulphuric acid. Many seaweeds contain a much larger proportion of sulphuric acid than this, and if, instead of escaping into the air, it were to come in contact with peroxide of iron, pyrites might well be formed.

The sulphur contained in the seaweed thrown up every year near Helsingors is enough to make 332,000 pounds of pyrites.

Probably, therefore, the large quantities of pyrites found in chalk are derived from seaweeds; and this, too, is likely to be the origin of the small percentage contained both in the blue limestone of the Jura (to which it is said to owe its colour) and in the bluish marl constantly deposited on the coast of St. Malo.

CHAPTER XIII.

ANIMAL SCAVENGERS—TERMITES, ETC.

Termites ; Great Strength of Nests—Rapidity with which they remove Dead Trees—Without the Termites there would be no Forests—Destruction of Furniture ; Houses and Deserted Towns—The Teredo removes Wrecks and Drift-wood, Ocean otherwise choked—Vegetable Matter brought down by the Ambernoh River—Food of Deep Sea Animals—Boring Mollusks—Wood-boring Beetles ; the Weevil and " Death-watch "—Caterpillars and Worms as Scavengers.

NATURE has other ways of disposing of her vegetable refuse besides those already mentioned.

In tropical regions, especially where the land is not fully cultivated, a fallen tree has but little chance of quietly mouldering away. Instead of this, it is devoured, and that with wonderful rapidity, by the various species of Termites, called " Bugga-bugs " in Africa, " Cupim " in Brazil, " wood ants " and " white ants " in the West Indies.

Ants they are not, however, for though, closely resembling them in many of their ways, they differ from them in the shape of their wings, &c., and their larvæ, or grubs, are active creatures, with six legs like the perfect insect, whereas the ant-grubs are legless and worm-shaped.

Termites belong to the same order of insects as the dragon-fly, that is, the Neuroptera, or nerve-winged order. In every nest there are three classes of insects—the labourers (Fig. 39), whose mouths are adapted for gnawing ;

the soldiers, who have large heads wherewith to inflict deadly wounds on an enemy; and the king and queen, who spend ·their lives in retirement, and are diligently waited on by their subjects. The queen is from three to six inches long, and lays about sixty eggs a minute, which are at once carried off by her attendants to the nurseries, there to be carefully watched and tended. The

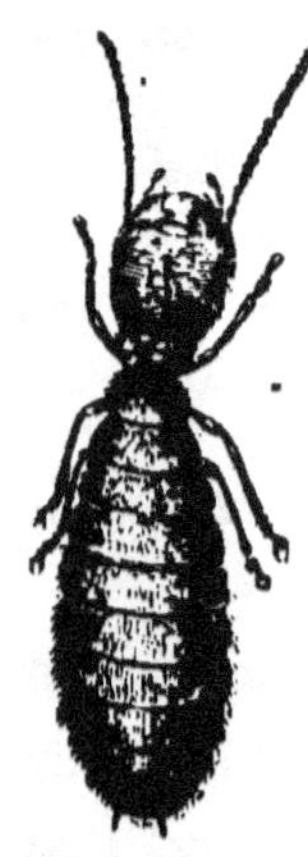

Fig. 39.—LA-BOURER TER-MITE.

nursery walls, by-the-bye, are slightly covered with "mould," which under the microscope resolves itself into minute white globules, of the size of a small pin's-head, and the shape of mushrooms; and it is probable these tiny fungi are grown purposely as food for the grubs.

The nest of the termite, surnamed *bellicosus* ("warlike,") is a wonderful construction, shaped like a hay-cock or sugar-loaf, and as it is from eight to twelve feet high, when several are built together they might easily be mistaken for a native African village. If our buildings bore the same proportion to our size as these nests do to the size of the termites, which are only about a quarter of an inch long,* we should be living in houses more than half a mile high, or nearly five times the height of the Great Pyramid. Yet they are completed in three or four years, and in the second and third year are covered with a growth of grass, &c., which, as it withers in the sun, gives them the appearance of haystacks. In Africa they are made of a yellow clay, which is

* The labourers are a quarter of an inch, the soldiers half an inch, long.

worked until perfectly smooth, and as it hardens becomes almost as solid as sandstone, and strong enough to bear the weight of a man or horse, and even a loaded cart.

The Cupim nests are smaller, but the walls are six inches thick, and so hard as to be cut open with difficulty.

In Ceylon the nests are not destroyed even by the monsoon-rains, which no mortar or cement is able long to withstand, and the clay is so extremely fine and pure that the goldsmiths there use it in preference to all other substances for the moulds of their finer castings. For the same reason it is used for making idols.

Other termites make their nests of black clay, and in the shape of cylinders, which, being three-quarters of a yard high, and having conical roofs with overhanging eaves, look like gigantic mushrooms. Others, again, use a sort of paste, made of wood, gum, and the juices of trees with which they build nests among the boughs as large as bushel-baskets, and strong enough to resist the fury of a tornado.

But, whatever their habitations, all the termites are alike in the wonderful rapidity with which they will remove anything perishable which comes in their way. It is said that not even fire and tornado equal in this respect the termite-hosts, which in a few weeks will destroy and carry away the trunks of large trees without leaving a particle behind. In this way they clear the ground for a fresh growth; and in the tropics, where vegetation matures rapidly, and the Guinea grass reaches a height of thirteen feet in five or six months, it is highly important that all plants should be removed as soon as they have reached perfection, and begin, as they do at once, to decay.

But for the termites there would not be a forest left in the world, for the dead would choke the living ; then the absence of foliage would so alter the climate that droughts would ensue, and the land would be turned into a desert ; so that, as Mr. Smeathman says, mankind would probably suffer less from the loss of one or two of the larger animals than it would from the extermination of the termites. They seldom attack a healthy tree, and probably when they appear to do so it will be found that disease of some sort has really begun its inroads ; but any stake in a hedge which has not taken root they at once destroy. If the bark be sound they enter at the bottom, and completely hollow it out ; otherwise they first cover it carefully with clay, as, though blind, they will not work in the light. It must be confessed that they do not distinguish as one could wish between wood that is useful and wood that is useless to man ; dead wood is never anything but dead wood to them, and thus they have been known to destroy all the timber-work of a spacious apartment in a few nights, so carefully concealing their ravages, however, that their presence could not be suspected.

In a single night in Japan they have made a tunnel as thick as a man's little finger through the floor, up one leg of a table, across the top, and down another leg.

Some large species begin work several feet below the foundations of a house, and tunnel their way up through the floors, and into the furniture, the position of which they seem to know with the utmost accuracy. The only way to prevent their ravages is to place the legs of every chair and table in pans of water, a precaution, which, of course, is out of the question with door-posts, &c., and thus it happens that all the timber of a house may be perforated in every direction until nothing but a thin crust, no thicker than a sheet of paper, remains.

In some cases they seem to know when a post has a weight to support, and are careful to fill it up with clay; but where the woodwork is not used as a prop, they see no need for such precautions, and the owner who is thus literally " eaten out of house and home " may be first made aware of the fact by finding his window-sills crumbling away beneath his touch.

Neither wine-casks nor scientific instruments are safe from them, and at Tobago they once caused the loss of almost a pipe of Madeira wine. But they do not confine their attentions to wood. An unfortunate engineer in Brazil who had just returned from an expedition with a collection of plans, &c., left his trunk on a table for the night and found next morning that all his clothes and papers had been destroyed, not a square inch of the latter being left, while every atom of pencil, lead and all, had entirely disappeared. Boots and shoes they will devour in a single night, and one wonders how the dwellers in equinoctial America contrive to keep a roof over their heads or clothes on their backs, and it certainly is no marvel to learn that it was a rare thing to find among the natives any papers more than fifty or sixty years old in Humboldt's day.

Several species inhabit the warmer parts of Europe, and a colony, probably imported from the West Indies, have established themselves in La Rochelle, where, besides doing other damage, they have eaten up the whole of the town archives, with the exception of the topmost sheet, under cover of which they pursued their destructive labours without attracting attention.

In America, where it is no uncommon thing to find wood-built towns and villages, once populous, suddenly, for one reason or another, abandoned and left to decay, the termites will come and take possession and completely clear the ground, doing their work, too, so thoroughly that not a door-post or a trace of one will be left, unless it happens to have been of teak or iron-wood. They will work, moreover, with such amazing rapidity, that in two or three years from the time of their arrival, the site of the town will be covered with a thick growth of vegetation.

In hot countries, however, Nature has a wonderful number of other labourers employed in the work of removing dead vegetable matter, though the termites are certainly the most expeditious of any.

What they are on land, that the Teredo (Fig. 40) is in the ocean, which, but for the labours of this creature, would be choked, in spite of its vastness, by the large quantities of timber, to say nothing of wrecks, which

Fig. 40.—The Ship Worm (*Teredo navalis.*)

it is constantly receiving. By far the larger part of this wood floats until it becomes waterlogged ; and this being the case, when it does at last sink, it must do so too far from shore to have much chance of being covered with mud or sand and converted into coal.

Seventy miles from the mouth of the Ambernoh River of New Guinea, the *Challenger* found the sea so blocked with drift-wood that her screw had to be constantly stopped. Long lines of it were passed, consisting partly of whole trees, but chiefly of broken pieces of wood with the stems of a large kind of cane-grass, various fruits and other fragments, and the seeds of inland plants, but no leaves, for these drop first near the shore. A wide area of the sea in this region is constantly covered with drift-wood, which seems to have a special population of its own. Much of the softer vegetable matter decays and dissolves and so helps to feed the whole kingdom of Protozoa as well as more highly organised forms of animal life. But the wood might last for ages under water, if there were no means of getting rid of it except by the slow process of decay ; for one of the old wooden ships of the Northmen has been lately dug up from the place where it sank and was buried some thousand or more years since, and oak is still in existence which is known to have been driven into the bed of the Thames in the time of Julius Cæsar, nearly two thousand years ago.

But neither wrecks nor drift-wood are left to accumulate, thanks mainly to the curious teredo or " ship-worm," which is really a bivalve having a very small shell, only a few lines broad in fact, while its greyish-white worm-like body is a foot long and half an inch thick. This creature bores deep tunnels in submerged timber, and does its work so quickly that a piece of hard sound wood is completely riddled in five or six weeks.

Though extremely useful in its right place, it has at times done much mischief; and it is as a protection against its ravages that ships are sheathed with copper, and the timbers of piers and jetties, &c., are studded with iron nails, the rust from which soon spreads over the whole surface and renders it unpalatable. Dockyards have sometimes suffered much from it; and the Dutch have been greatly alarmed by its attacks on the wooden piles supporting the all-important dykes which alone preserve their country from being flooded by the North Sea.

The pholas, rock-boring Venus, &c., already mentioned as piercing stone, also help to clear away dead wood.

But to return to the land, where there are numerous other removers of dead vegetable matter. In the tropics a large proportion of insects of all orders, but especially beetles, are more or less dependent upon vegetable matter, particularly bark, timber, and leaves in various stages of decay, and the number and variety of insects which may be collected in a given time depends upon the number of trees which have been or are being cut down. In the Aru Islands, we are told by Mr. A. R. Wallace, no sooner is a tree felled than it is attacked by swarms of little wood-borers, hundreds of which perish from their over-eagerness, being glued into their holes by the outflow of sap.

Numerous species of small beetles lay their eggs in dead wood, and the larvæ, as soon as hatched, begin making tiny galleries in all directions. A quantity of oak timber was thus once destroyed in the royal dockyard of Sweden, and we all know the look of old furniture which we call "worm-eaten." Each of the small holes is in fact the door of a gallery made by some beetle-grub.

Some years since a whole cargo of cork was destroyed by minute beetles and their grubs, which also damaged the timbers of the ship in which it was conveyed. Much loss has also been occasioned by beetles which have chosen the corks of wine-bottles as a convenient place to lay their eggs in.

The most destructive wood-eating beetle in England is the weevil, which gnaws a hole in the bark and then drives a tunnel into the solid wood, where it lays its eggs and frequently dies, thus effectually stopping the entrance with its own body against all enemies. The eggs are soon hatched and then each grub burrows a tunnel for itself from an inch and a half to two inches long, which it widens as it grows. A number of these insects together will entirely peel a tree, and thus cause, or probably rather hasten, its death, since it seems very doubtful whether they ever attack a perfectly healthy tree.

It would take too long to mention separately all the various beetles which feed on decaying vegetable matter, wood, leaves, bark, seaweed, and fungi. Fungi are frequented by very many species, which, as they are not found elsewhere, seem to be entirely dependent upon them for food. It may, however, be stated that the ticking sound, called the "death-watch," which is often heard in old houses, is the call of a small beetle always found in dead wood.

In cold climates the caterpillars of the wood-leopard

(Fig. 41) and goat-moths feed on decaying timber, and as the latter spend about four years in the caterpillar state, apparently never ceasing to eat, they bore considerable tunnels not only in fallen trees, but in those which, but for them, might long remain standing.　The goat-moth caterpillar when full grown is three inches or more long and as thick as a man's finger ; and since as many as sixty-seven have been found

Fig. 41.—THE WOOD-LEOPARD MOTH.

in the mere splitting of a piece of trunk two feet long, it is no wonder that many an elm and willow are unable to stand against the wind after being colonised by them.

It is perhaps when we consider the vast amount of damage which wood-eating insects inflict upon us, that we best realise the great services they also render, for in a fully cultivated country it is the havoc which they play that chiefly comes under our notice.

Certain fly-grubs also feed on rotten wood, decaying roots, &c. ; the jet ant makes its nest in decaying trees, and so do the wood-wasps which abound throughout Europe and North America.

The titmouse and woodpecker are sometimes accused of injuring trees, but it seems unjustly, for they are said never to bore into healthy bark. There are grubs beneath, however well they may look, and it is for these they tap the wood. At the same time, when a tree is diseased, they help on the mischief by making a hole in the tainted wood, where they build their nests and open a way for the rain to penetrate.

Worms may be said to be very perfect scavengers as far as their powers go, since they not only remove refuse, but turn it to good account as manure, and that without rendering themselves disagreeable, which is more than can be said for some other creatures. They are omnivorous, and besides dragging large quantities of leaves down into their burrows, as linings as well as food, they also feed on decayed flowers, even their own dead comrades, and in fact decaying matter of all kinds. They add largely to the organic matter of the soil, and therefore to its fertility, not only by the enormous quantity of leaves they carry down, but by burying bones, shells, leaves, twigs, and refuse of all kinds beneath their castings. The leaves they feed upon are torn into small shreds, partially digested and mixed with earth, and it is this which gives vegetable mould its dark tint.

Slugs and snails hardly deserve any notice as scavengers, since, though they do eat fallen leaves, they live chiefly on sound ones, as the gardener knows to his cost.

CHAPTER XIV.

SCAVENGERS—ANTS, FLIES, AND BEETLES.

Ants employed by Naturalists, Destruction of "Specimens," Black Ants, Red Ants, Fire Ants—Ants bury their dead—"Horse Ants," "Travelling Ants"—Extermination of Vermin, difficulty of guarding against Fire-ants, Mushroom-growing Ants—Scavengers kept by Ants and Spiders—Three Flies equal to a Lion—Keen Sense of Smell—Beetles feed on Animal and Vegetable Matter—Burying Beetles—Carrion Beetles — Skin-eaters — Skeleton-makers — Cock-tails — Scarabæus — Dumbledor Moths.

AMONG the most valuable natural scavengers of Ceylon are the ants, for they never sleep, work night and day, and remove every particle of decaying or putrid matter in a marvellously short time.

They are often turned to account by the naturalist, who gives them his shells to clean, and finds that in a few days they remove every vestige of the dead mollusk, even from the innermost whorls and recesses, which he could not himself by any means reach. A bird, or other small animal, if buried near an ants' nest, in a box pierced with a few holes, will speedily be converted into a perfect and most delicately whitened skeleton by these industrious creatures.

The said naturalist may, however, have reason to complain bitterly of his little servants at times, for they are not discriminating, and unless he be on his guard against them, he may find his most valuable collections of insects either totally destroyed or cut up into pieces of a convenient size for removal. Ants have a special affection for insect "specimens," and the only way to protect these is to keep the legs of the table on which they are placed in pans of water. Even this does not always answer, as a thin film of dust will make a floating bridge strong enough to bear the weight of the smaller species. Against oil, however, they are all alike powerless.

Mr. Wallace mentions that in New Guinea, a small black ant took possession of his house, built nests in the roof, made covered ways down the posts and across the floor, and also occupied the boards he used for pinning out his butterflies, filling up the grooves with cells and storing them with small spiders.

The red ants which in the Moluccas frequent houses are " a most terrible pest," for " they form colonies underground and work their way up through the floors, devouring everything eatable. It is very difficult to preserve bird-skins or other specimens of natural history where these ants abound, as they gnaw away the skin round the eyes and the base of the bill; and if a specimen is laid down for even half an hour in an unprotected place it will be ruined."

"I remember once," says Mr. Wallace, "entering a native house to rest and eat my lunch ; and having a large tin collecting-box full of rare butterfles and other insects, I laid it down on the bench by my side. On leaving the house, I noticed some ants on it, and on opening the box found only a mass of detached wings and bodies, the latter in process of being devoured by fire-ants," as some of these red ants are called from the extreme sharpness of their sting.

But then how are the ants to know when a dead body is wanted and when not, since their own opinion clearly is that the dead should never be allowed to remain among the living?

They alone of all animals, so far as is known, are constantly in the habit of removing and burying their own dead, and, indeed, have their regular cemeteries; but, while thus careful of their friends, they suck the juices of strangers and enemies and throw the dry husks together, in some spot away from their nest. Their slaves they likewise bury, but in a place apart, and are seldom known to eat the dead bodies of either slaves or comrades.

Mrs. Lewis Hutton, of Sydney, gives a most curious account of the funeral customs of some Australian soldier ants. Having killed several ants which had attacked her child, she saw their friends come and carry them off in procession, each body being followed by ants which she took for mourners. These mourners were then apparently called upon to give their help in filling in the graves, which some did, while six or seven who refused were killed, and buried without honour in a single grave.

In our climate ants are essentially carnivorous. A company of horse- or hill-ants have been seen dragging away half a dead snake of the size of a goose-quill, and no doubt they do much useful work which escapes notice; but it is in the tropics, where they are omnivorous, that the work they accomplish is best appreciated.

The "travelling ants" of South America start on their periodical journeys just before the rains set in, and not only clear every bush and low tree in their path of all insect-life, but also enter houses. Their sudden arrival is often announced by the scuttling across the floor of some alarmed cockroach, pursued by one small ant, which does not look like a formidable enemy. But the cockroach knows better, for the one little ant is but the forerunner of an enormous army, and very soon three or four others appear and join in the pursuit, and the fate of the cockroach is sealed.

When a house is thus invaded, the rightful owners can do nothing but give it up for a time to their uninvited guests; unwelcome we can hardly call them, because in hot countries many creatures, which pay neither rent nor taxes, are in the habit of establishing themselves in all parts of the house, and now at the first notice of the approach of the ants, away they all rush as fast as they can go.

A lady who has lived for years in Trinidad, says that the arrival of the ants is hailed with delight, for they investigate every corner and crevice in the house, the walls, ceilings, and floors being black with their countless legions; but when they have thoroughly explored the premises, which takes but an hour or two, they leave them quite cleared of all living things which have no business there. Rats, mice, snakes, cockroaches, spiders, scorpions, and even the fleas, have vanished for at least one or two months, and the lawful owners can live in peace.

The ants themselves decamp when their work is done, or should they linger may be easily dismissed, it is said, by a little cold water. The driver ants of Africa, which also enter and clear houses, are reported to kill even the great python; and the ants of the American plains are employed,

o

according to Mr. McCook, as vermin-killers by the Indians, who spread their furs and blankets upon or near an ant-hill and soon find them perfectly cleared of eggs, larvæ, and insects. Strictly speaking, however, it is not when they prey upon the living, but when they keep to their usual diet of dead insects, that these American ants can be regarded as scavengers.

The small red fire-ant, Mr. Bates says, though found in most open places along the banks of the Amazons, wherever the soil is sandy, seems to have its head-quarters at Aveyros, a village on the tributary Tapajos, which is completely undermined by its galleries. The Tapajos is nearly free from the insect-pests of other parts, a fact which may perhaps be put to the credit of the fire-ant, but the latter seems to be as great a plague as all the rest together. At one time Aveyros was actually deserted in consequence of their attacks, and though their numbers were supposed to be diminished at the time of Mr. Bates's visit, the houses were still overrun with them, they disputed every fragment of food with the inhabitants, and destroyed clothing for the sake of the starch. Eatables had to be suspended from the rafters in baskets, the cords being well soaked in a kind of balsam. So malignant and unprovoked are their attacks that the cords of hammocks must be smeared with the same balsam to keep them off at night, and those who would enjoy an open-air chat with their neighbours in the evening, must rest their feet on stools and sit on chairs the legs of which have likewise been well smeared.

At Ega on the Upper Amazons, Mr. Bates had to keep his specimens in cages suspended from the rafters by cords well anointed with a bitter vegetable oil to preserve them from the attacks of the ants.

A curious fact has been observed with regard to the mound-building ants, whose nests contain certain chambers which appear to be lumber rooms, some of them being filled to the roof with husks, hulls, and decayed or mouldy seeds, and then sealed up, presumably in order that the contents might not be unpleasant to the inmates. Others are filled with gravel, which Mr. McCook believes to be what remains over after the roofing of the mounds. For these ants invariably cover the top of their habitations, which are sometimes thirty or forty yards round and from one to three feet high; and the covering consists of sand, gravel, small pebbles, with little bits of limestone, fossils, coal, gold-dust, fragments of valuable minerals, whatever, in fact, they may happen to bring up in the course of their excavations. Ants in India have been observed to ornament their nests with garnets from the sea-sand; and their relations in England carry off bits of amber-like resin, whether for food or decoration, we cannot say; but there seems no reason whatever why they should not like to have pretty things or even curiosities about them, since they are so extremely human in other respects.

Another inhabitant of South America, the bizcacha, which is like a large rabbit, collects bones, stones, thistle-stalks, and hard lumps of earth, round the mouth of its burrow, and has been known to add a gentleman's watch to its treasures; while the satin birds of Australia erect bower-like structures of twigs and branches, which they adorn with coloured feathers, bones, and shells. These curious covered arcades are sometimes several feet long, and seem to serve as a common pleasure-ground for a number of birds, which amuse themselves by running in and out.*   The mere size of the ant is of course no reason at all against its taking pleasure in similar "collections."

Sir John Lubbock is at all events of opinion that they may take pleasure in keeping pets, for as many as forty different species of minute beetles are found in their nests, some of which are so thoroughly domesticated as never to be met with elsewhere.

The ants take great care of these beetles, and are as anxious to remove them as their own young to a place of safety should the nest be invaded. But, as beetles are the great insect-scavengers, it seems probable that some at least may act in this capacity to their hosts, who keep a blind woodlouse for the same purpose apparently. The latter, however, they treat with the utmost unconcern, and leave behind to shift for itself when they migrate.

According to Mr. Belt, the Saüba, or leaf-cutting ants, like the termites, are in the habit of growing "mushrooms," but in a still more systematic way, since they supply them with the proper soil or manure, in the shape of leaves, the blossoms of certain plants, and the inside white rind of oranges, all of which are torn into minute shreds. Certain it is that some of their chambers are often three parts filled

* The jackass-penguin, according to Mr. Moseley, collects in and about its nest small stones, shells, old bits of wood, nails, rope-ends, old sails, boat-spars, even forgotten bags of guano, and anything else which may chance to be left in its neighbourhood, but this seems to be for purposes of drainage.

with a speckled brown spongy-looking mass, which, on close examination, is seen to consist of tiny bits of withered leaves, overgrown by a very small white fungus ; and should the nest be disturbed, the ants are evidently most anxious to carry every morsel of this " food " under shelter, and when they migrate invariably take it with them. It is also certain that they do not eat the leaves themselves, so one can only conclude that they need them for the purposes of cultivation. The refuse-particles, when exhausted as manure, are stowed away in deserted chambers, and serve as food for the grubs of beetles.

Rose-beetle grubs are often found in the nest of the wood-ant (Fig. 42), probably for the sake of the chips of wood, fir-needles, &c., which they need for their cocoons.

Nor are ants the only creatures which keep scavengers. Some of the larger spiders allow certain small species to live on the outskirts of their large, strong webs, and to feed on such minute insects as are beneath their own notice, and would otherwise not only be wasted,

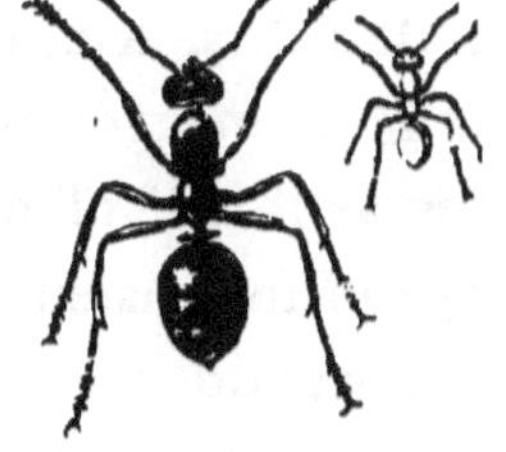

Fig. 42. — WORKER WOOD-ANT.

but become a nuisance by choking up the web. Mr. Darwin tells us that the booby and noddy of the desolate St. Paul's Rocks, where nothing in the shape of a plant, not even a lichen, grows, have their attendant scavengers, in the shape of a woodlouse, a feather-feeding moth, and a beetle.

Among the most important scavengers we must reckon the flies, since there are very few which do not, at one time or other of their existence, feed on decaying matter, animal or vegetable. Flies are the first to attack any dead body which may be left exposed, and, by perforating it in all directions, they open and prepare the way for the host of beetles which follow them.

Linnæus, the great naturalist, declared that three flies were equal to a lion as regards the length of time they would take to demolish the carcase of a horse; and this will not seem by any means impossible when we learn that each flesh-fly produces 16,000 to 20,000 maggots, which are hatched in her body, and are therefore ready to begin feeding at once, which they do so voraciously that in many species they increase their weight two-hundredfold in twenty-four hours.

The larval state of all insects, whether butterflies, moths, beetles, flies, &c., is the grand feeding-time, when, indeed, the grub, maggot, or caterpillar, does nothing but eat, and as soon as it is "full fed" prepares at once to enter on the next stage of its existence.

Flies do not wait for decomposition to set in before coming to feed and deposit their eggs or young; or rather, perhaps, since decay actually begins as soon as life ceases, we should say that their senses are keener than ours in detecting it. A mouse has been known to be full of their maggots two or three hours after death; and every butcher knows to his cost how attractive his meat is, even when freshly killed. Human beings endowed with the ordinary sense of smell are able, it is said, to detect the three hundred millionth part of a grain of musk; but insects require a very much smaller degree of odour to guide them to their food. Thus it has been noticed in Ceylon that, in a few moments, or even less, from its death, an elephant will be covered with myriads of black flies, not one of which had been visible before.

Not only can no odour whatever be detected by human noses, but the sudden arrival of the flies is often the first intimation to the bystanders that the animal is really dead; yet some extraordinarily subtle odour there must be, and it must have travelled through the air with a speed which is equally extraordinary.

White sugar is to us utterly without smell, yet Sir J. Emerson Tennent mentions that the smallest particle, though wrapped in paper and placed in the centre of a table, is quite enough to attract the small black ants, a line of which will be formed in a few minutes to effect the removal of the delicious morsel.

Flies seem to be guided by scent and scent only; for, when she cannot get at the meat, the bluebottle, or blow-fly, will lay her eggs on the wire-gauze meat-cover, and in this case one would say she must be deceived by the smell, for she can hardly know that, though it is beyond her own reach, her eggs will fall through and her children thus find their necessary food. It is, at all events, quite certain that gauze is no sufficient protection against her and her family.

In tropical countries where the extreme heat and dryness of the air combine to shrivel up any dead body so rapidly that it can hardly be said to putrefy, and so completely that travellers in the woodless pampas can make their fire of a dead horse, carrion-beetles, though not absent, are not numerous. On the other hand, they abound in damp temperate climates, where the slow rate at which decomposition proceeds makes their services especially valuable.

Many feed both on animal and vegetable matter, especi-

Fig. 43.—Burying Beetles.

ally on fungi in a state of decay, and their numbers are so vast that we can mention only a few of the most active. Foremost among them are the Sexton, or burying beetles, whose scent is so keen that they come from great distances to find their food. (Fig. 43.)

In the vicinity of towns they feed on all manner of garbage, hunting in couples and chiefly at night, the male insect wheeling round and round in the air like an eagle before pouncing on his prey. After a careful examination of their booty (perhaps a dead bird), they proceed to make a hearty meal, and then explore the immediate neighbourhood for a spot where the ground is soft enough in which to bury the remainder. This found, they drag the body with much labour to the place, and the male insect begins operations by digging a furrow all round the body at the distance of half an inch, using his head as a spade. Another furrow is dug within the outer one, and so on, until after several hours' most laborious work, during which the insect has been obliged at times to rest from sheer fatigue, he at length goes underneath the body and pulls it by its feathers into the hole scraped beneath. His wife, meanwhile, has been quietly seated among the bird's feathers, and now allows herself to be buried with it. Her husband treads down the body, shovels back the earth, treads it well in, scrutinises it carefully to make sure all is right, then makes a hole in the still loose earth and, burying himself, rejoins his wife. The great object of all this hard labour has been to secure a proper place for her in which to lay her eggs, which she does in a number proportioned to the size of the bird, after which the two creep out and fly away. The great business of the larvæ, as we have said, is to eat, and the parents are careful to provide them with food enough, though not too much, to last them until they are full grown. And the larvæ do their work well, wasting nothing, but consuming even the skin and bones.

Sometimes, so Miss Stavely tells us, the Sexton beetle makes a hole nearly a foot deep to receive the carrion, and will bury creatures many times larger than himself, such as birds, frogs, and even rabbits. If the body is too large for one, others come to help and to feast with him. In Ceylon, one beetle has been seen to bury a mole forty times its own weight, without assistance, and four together will bury a crow. The Phanæus, a beetle one inch and three-quarters long, with one large horn, buries dead snakes in a few hours.

Mr. Westwood mentions that in the course of fifty days he has known four beetles to bury four frogs, three small birds, two fishes, one mole, two grasshoppers, the entrails of a fish, and two pieces of ox liver. It is, indeed, chiefly owing to their burying habits that we may wander for hours in the woods or fields without seeing such a thing as a dead bird, mouse, rat, &c.

Some of the burying beetles work in company, others alone, and while some, as we have seen, dig with their broad flat heads, others do so with their fore legs. Their scent is so keen that they can detect their food from a wonderful distance, and they have been known to undermine the stick to which a dead mole was fastened, in order to bring the dainty morsel within their reach.

Besides the Sextons, there are in England alone many hundred species of beetles which feed on carrion without burying it, and, as soon as the flies have opened the way, they arrive in hosts, accompanied by wasps, hornets, and ants, the last relics of the feast being consumed by a tribe of small beetles, so that nothing is wasted.

Carrion-feeding beetles, like other larger carrion-feeders, though invaluable in the service they render, are themselves decidedly unpleasant, owing to their extremely strong and disgusting scent. The hands will smell for hours, and the coat for days, after being in contact with them, unless, indeed, they happen to have been fasting for some time from their highly-seasoned food, when they are quite free from smell.

Many of the carrion-feeders prey upon the living, as well as the dead, and, as before said, eat decayed fungi and other vegetable matter. Some lie in wait on the banks of rivers, and devour the dead dogs, cats, &c., thrown up by the water, as well as small mollusks, alive or dead. They will, in fact, clean your shells and skeletons as well as the ants; but, with all their usefulness, grievous are the complaints made against some of the race, especially those known as skin-eaters (*Dermestes*).

These render infinite service, as we are assured, but, at the same time, they and their minute grubs will entirely destroy your books, eat up your furs and your natural-history collections, besides invading your larder, and feasting upon the dried meat, bacon, &c. So injurious were they some years ago in the large skin warehouses in London that a reward of £20,000 was offered, and offered in vain, to any one who could devise effectual means of preventing their ravages. They are commonly found in the bodies of moles stuck up in the fields to dry, and will consume not only the flesh but the skeleton; nothing, in fact, comes amiss to them, for they will devour horns, hoofs, and leather, even in the form of old shoes. (Fig. 44.)

Mr. Buckland says that he has never found a dry body of an animal that had not in and about it specimens of the *Dermestes lardarius,* or bacon-beetle, called a "hopper"* by the ham and bacon merchants.

"They are," he writes, "capital skeleton makers, and if the skins of the creatures in the gamekeeper's museum

Fig. 44.—SKIN-EATERS.

be removed, the skeletons will be found underneath in a most perfect state of preservation, and quite fit, after a little washing, for the cabinet. The late Mr. Baker, of Bridgwater, took advantage of their powers by setting them to work to make skeletons of delicate things, such as small birds, fishes, frogs, lizards, &c. Neat workmen are these little hoppers, touching nothing but the flesh, and they clean much better than ants. The animal to

* "Hoppers" are the larvæ, or grubs.

be made into a skeleton should be soaked in water, to get all the blood out, then dried and placed with the hoppers in a covered box; a few birds' feathers should be put over them, as they will work only in the dark."

Another great family of beetles is well known by their

Fig. 45.—THE DEVIL'S COACH-HORSE.

peculiar habit of turning their tails over their backs, after the manner of scorpions, which has earned for them the name of Cock-tails. They are also called Rove-beetles, from their great agility, and Short-wings, because their outer, or sheath-wings, are so short that the long wings they use in flying need packing and arranging before they can be tucked away underneath. This they do with the tip of their tails, and hence are obliged to turn them up over their backs; but the movement seems also to be made in self-defence, possibly with an idea of inspiring terror. And the Devil's Coach-horse, one of the largest and commonest species, more than an inch long, certainly is a repulsive and ferocious-looking insect. (Fig. 45.)

Most of the little black "flies," which annoy us by getting into our eyes, are really minute Cock-tails, some no thicker than a horse-hair; and much of the irritation they cause is due to this habit of turning up their tails the instant they alight.

There are about a thousand species of Cock-tails, and all are extremely voracious, and all, more or less, scavengers. One, called the "fish-fly," is most unpleasantly abundant on the shores of Newfoundland, where it feeds on the dead and dying cod; but they do not confine themselves entirely to carrion, and some of them are extremely ferocious, not only hunting their prey, but attacking and devouring their own kind. Many, especially the Coach-horse, are inveterate insect-feeders, and as such ought to be looked upon with favour by the gardener.

Dung-beetles, of which there are many hundred species, are especially numerous in the tropics, where animal life is most abundant; and first among them is the great scarabæus tribe, species of which are found in all the warm parts of the world. The sacred Scarabæus of Egypt (Fig. 46), which is hornless, is common in the south of Europe, and throughout Africa. It has a very odd appearance when walking, as its legs are set far apart; but it is this peculiar structure which enables it to roll along a ball of dung from an inch and a half to two inches in diameter, which it does standing almost on its head, and pushing the ball with its hind legs, much as a horse backs a cart. The earth being usually hard and stony in the countries where

Fig. 46.—THE SACRED SCARA-BÆUS.

the Scarabæus chiefly dwells, it has to search for a soft spot in which to bury the ball, which is exactly proportioned in size to the number of larvæ it will have to feed. The labour of rolling it is great, for it is never quite round; but the persevering beetle toils on with wonderful determination, though obliged now and then to stop and rest before it can continue its exertions. Tired or not, however, nothing will induce it to give up the precious ball in which it has laid its eggs, though it will be equally well satisfied with the ball of a neighbour, and shows no other mark of attachment to its offspring, which when once buried with their supply of food, are left to care for themselves, as they are well able to do.

Black is the usual colour of the various species of Scarabæus, but some are resplendent with the richest metallic colours. Some of the species add carrion to their other diet, and feed on dead fishes, besides acting the part of sexton to dead snakes, which they speedily bury.

There is no English Scarabæus, but we have many beetles which are similarly useful, some being especially plentiful on turnpike roads, and others in pasture fields. The earth-borers, as one large family of various species is called, dig with their fore-legs, which are especially adapted for the purpose, being powerful and notched.

Though hardly to be seen in hot, dry weather, the rain has no sooner softened the ground sufficiently for their operations than they appear in swarms, and work with astonishing rapidity, both clearing away what is offensive to sight and smell, and turning it to account as manure for the fields, thus doing a double service, which is of the greatest value.

Of all the English species not one is more useful and active than the heavy, unwieldy Dumble Dor, Clock, or Flying Watchman, which is often to be seen lying helplessly on its back. The Dor is usually of a glossy violet or blue-black, sometimes greenish-black, and sometimes quite black ; but, whatever its colour, it is always glossy, and always clean, in spite of the very unclean nature of its work, in which, as in all other respects, it may be considered a model scavenger, though how it can escape unsoiled is a mystery.

It is sometimes found in decaying fungi, but it is usually attendant on cows, and Mr. Wood mentions a pasture field which was entirely cleaned by it in the course of three or four days, the ground being literally riddled with as many as forty or fifty burrows in each square foot.

Unlike the Scarabæus, the Dumble Dor has no need to spend time and strength in search for a convenient spot for its operations. It merely digs through the patch which is to be removed, carrying some with it to the bottom of the hole, which is as many as twelve inches deep. It then

Fig. 47.—The Clothes Moth.

lays one egg, crawls up to the top, and sets to work again on another burrow—a truly gigantic work when one considers the size of the workman employed.

A vast army must have been engaged in clearing the field Mr. Wood describes, for it would, as he says, have occupied a strong body of men a considerable time, and then, after all, they could not have manured it as the beetles did, without first removing the turf.

Has it ever occurred to us to consider what becomes of

P

the old birds' nests? Were they left to accumulate from year to year, the trees would be so clogged with them that, as Mr. Wood points out, they would be unable to put forth their leaves, and must therefore die. Beetles and moths, however, come to their aid, and by devouring the sheep's wool, feathers, &c., with which the nests are lined, make it easy for wind and rain to scatter the other materials.

Two species of moths are especially given to attacking old, greasy clothing, such as horse-rugs, which have been left untouched for a few days (Fig. 47). Others turn their attention to carpets and carriage-linings, and another, whose proper food is also wool, will eat hair of any kind, even horse-hair, and will shave the fur off a skin more neatly than a razor.

"The *Tineida epigraphia* is the smallest of all moths," says Alphonse Karr, "being two lines wide, when its wings are outspread; but how magnificently it is attired! It is robed in gold and silver, and on the silvery gauze of its upper wings is traced, in letters of gold, an inscription which no one has yet succeeded in deciphering, though I fancy I can read it thus: *Maximus in minimis Deus* (God is greatest in His smallest works). In their larval state these moths eat furniture, silk dresses, furs, and even make so bold as to attack the fur-caps of the grenadiers."

CHAPTER XV.

SCAVENGERS—CRUSTACEANS, BIRDS, AND MAMMALS.

Protozoa—Crustacea : Sand-hoppers, Shrimps, Crabs—The Vultures of the Ocean—Vultures on Land: "Daddy Long-beard," "Pharaoh's Hens," Turkey-buzzard, King of the Vultures—Crows—The Gallinazo, protected by law, its great boldness—The Caracara and Arabian Kite—The Adjutant, Stork, and Domestic Duck—Rats: their great usefulness, personal cleanliness, fondness for Ivory; how they have been made use of in Paris—Pariah Dogs, Dogs of Constantinople, Jackals, Wolves, Hyænas—Nature's Scavengers not perfect.

WHAT becomes of the dead bodies of the Polar bears, reindeer, birds, &c., which exist in thousands and millions, and must die in untold numbers?" Professor Nordenskjöld asks the question, but does not answer it, though he says that "self-dead" animals are so rarely seen in the Arctic regions, that it is easier to obtain fossil bones than recent ones. Evidently, therefore, the scavenging is very thoroughly done by some means or other.

Mr. Bates is equally struck with its thoroughness in another department, namely in the fresh-water pools or lakes on the Upper Amazons. "How elaborate must be the natural processes of self-purification in these teeming waters," he says, "for the water is quite pure, no scum of confervæ, or trace of animal decomposition is to be seen on the surface, no foul smell is ever perceptible, and the whole of the level land is most healthy, instead of being covered with malaria-breeding swamps."

Much of this scavenging is no doubt accomplished, as has already been said, by vegetation, and much also by the varicus Protozoa, multitudes of which exist everywhere in still water.

But, to go on to creatures whose labours we are better able to appreciate, among the most active coast-scavengers must be reckoned the great army of the Crustacea, all of which are flesh-eaters, from the water-flea to the lobster, and all, it would seem, voracious, though some are more particular than others.

The sand-hopper, for instance, will feed on almost any-thing which is soft and capable of decay, and does not despise a meal of seaweed; the crab will eat only fresh food, and putrid bait is reserved by the fisherman for the lobster, which is said to prefer its food highly seasoned.

The thousands of little creatures, exactly like common woodlice, which swarm about cliffs and piers, are vegetable feeders, and will devour, according to Frank Buckland, the planks of boats and eat up sails and nets if these are left undisturbed for any time; but the sea-lice eat animal food, and "in the United Service Museum are some very perfect skeletons of sea birds," made by them in the Arctic regions. "The birds were let down into the sea to an immense depth, and left there twenty-four hours; these bones are as white as ivory."

Though they do not come under the head of crustaceans, we must here mention that tadpoles are likewise good skeleton-makers, and that very perfect skeletons of small animals are sometimes found made by them in ponds. Tadpoles will eat up a dead kitten, and will also condescend to decayed vegetable matter, but when other food fails they turn cannibals and prey upon one another.

Among the lesser fresh-water scavengers are certain little long, narrow worms, with which dead, and even sickly fishes are often found to be covered.

But to return to the crustaceans, among which the common shrimp (Fig. 48) is pre-eminent as a "capital skeleton-

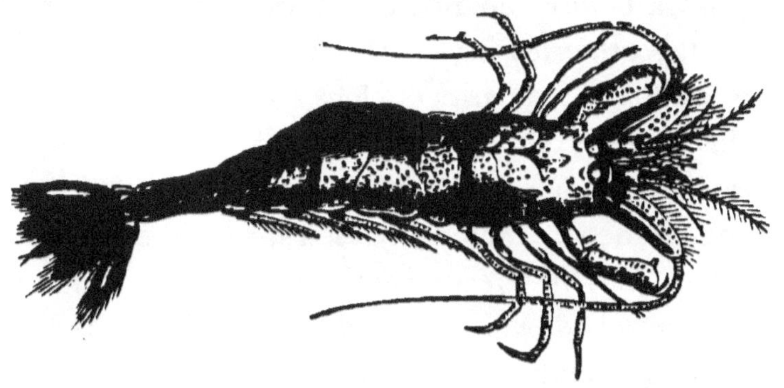

Fig. 48.—THE COMMON BROWN SHRIMP.

maker." So voracious is he that he is called, *par excellence,* the "scavenger of the ocean," but the crab, both in water and on land is an equally diligent consumer of dead animals; and among crabs one of the most useful is the Thorn-back Spider-crab, which, though he does good service in the sea, is more especially useful along the coast, thanks to his large appetite and the keen sense which guides him unerringly to his food. But for his labours the sea-shore would often be anything but a charming recreation ground, for the sands would be strewn with the dead bodies of unsaleable fish, as well as with the offal resulting from the processes of skinning and dressing often performed there by fishermen. As it is, however, the crabs arrive in such myriads and work so busily that by the next tide hardly a trace of anything unpleasant remains, though bones picked perfectly clean may be seen strewing the sand in numbers, showing how abundant the banquet has been.

The Spider-crab, assisted by others of different species, sets to work boldly, holding on to the fish with one claw and with the other tearing off the flesh and conveying it to its mouth with the regularity of clockwork, or of a Chinaman plying his chopsticks. So strong and sharp are its claws that no muscle is tough enough to withstand them, and the fish bones are cleaned as thoroughly as if they had been scraped by a knife. But in the process, a number of minute frag-ments are detached, which, though too small for the large pincers of the crab, would yet make their presence very unpleasantly felt if they were left to decay. This, however, is not the fate of these tiny scraps, for the crab carries with it a whole army of lesser scavengers, for which this mince-meat is exactly adapted. These are various small zoophytes, such as the sea-fir, &c., which attach themselves to its armour, sometimes in such numbers as to entirely cover both body and limbs, and thus are always on the spot to clear up the relics of the feast. Crabs will devour any kind of animal food that comes in their way, and on the desolate St. Paul's Rocks, which are tenanted only by countless multitudes of sea-birds, they quickly pick the bones of any dead individual.

Eels also feed on carrion, and most fishes are as ready to eat the dead as the living, so that any carcass which is washed into the sea, or even left near shore, is soon disposed of.

The large white-winged Glaucous gull, which builds in Greenland and Iceland, and comes as far south as Shetland in the winter, resorts to the entrances of the more exposed bays, or waits about a few miles from shore in attendance on the fishing-boats, to pick up the offal thrown out of them. The Dutch call it the Burgomaster or Chief Magistrate of the birds in Spitzbergen, where it follows the whale-fishers, in company with the Fulmar Petrel and Kittiwake, or Ivory gull, which it forces to give up their most dainty morsels whenever it takes a fancy to them. It is very dexterous in carrying off its food on the wing.

The common gull may frequently be seen hunting for refuse on any sandy flats, such as those about the Thames, but the Skua gull is seldom seen so far south. It also feeds on dead whales and other carrion, but it is a bird of low and disgusting character, justly called the parasite, for it is supported by the labour of others, and harasses the smaller gulls, which it obliges to disgorge their prey to satisfy it.

The Fulmar Petrels follow in the wake of the whalers as soon as they have passed the Shetland Isles, being very greedy of whale-fat, and as soon as a whale is cut up they flock together in thousands and follow the boats boldly at the distance of but a few yards. When carrion is scarce they follow the living whale, and so point out his whereabouts to the fishermen ; but should a dead whale chance to be stranded anywhere, they cannot make much impression on the carcass till some more powerful bills have torn away the skin.

In New Jersey, the black-headed gull haunts the neighbourhood of farmhouses and river banks, picking up garbage and other refuse of the fishermen ; and about the middle of May great multitudes assemble in the Delaware Bay to feed on the remains of the king-crabs left by the hogs.

Gulls (Fig. 49) are the vultures of the ocean, and besides keeping careful watch over the labours of whalers and fishermen may be seen in numbers following a shoal of porpoises and picking up the bitten and wounded fishes, which the porpoises themselves have not time to do when going full chase after their prey.

Vultures, though chiefly frequenting warm countries, are found more or less all over the world, and, though most of them are disgusting in their ways and unpleasant both to sight and smell, are almost everywhere protected in civilised countries for the sake of the great services they render.

In the south of Europe they are kept in the market-places, as storks are in Holland, and for the same purpose—to eat up the garbage ; and, being protected by heavy penalties, are extremely familiar and independent ; and at Natchez swarm in such numbers that all the refuse of the place is not enough to feed them.

The vultures, indeed, do their scavenging on a more extensive scale than either the dogs of Constantinople or the storks of India and Holland.

The largest of all the vultures are the Great Bearded

Fig. 49.—THE GREAT BLACK-
BACKED GULL.

Vulture or Lämmergeier, and the great Condor of the Andes, about which so many exaggerated stories have been told. They are of about equal size, averaging from eight to nine feet across the wings ; both seem to prefer carrion, and are invaluable scavengers, but will kill their prey when necessary, and are looked upon as foes by shepherds and herdsmen. The favourite haunts of the Condor are the regions of perpetual snow, from which it seldom descends; and the Lämmergeier frequents the Alps and higher mountains of Europe, a variety of it being also found in many parts of Africa, where it is called "Daddy Long-Beard." Bruce, the traveller, mentions that on one occasion when he was cooking his dinner on a mountain top, one of these birds boldly swooped down and put its foot into the pot where some goat's flesh was boiling, but not being prepared for the heat, withdrew it again speedily. Being very fearless, however, it at last managed to carry off a leg and a shoulder from the dish. When shot it was found to measure eight feet four inches across the wings.

The Egyptian or Alpine vulture, which inhabits Europe, Asia, and Africa, being nearly white, is called "White Crow" by the Dutch, and "White Father" by the Turks. A pair are attached to every group of natives in South Africa, and are to a certain extent domesticated, being perfectly harmless, and very useful in clearing the premises of offal. In the East they walk fearlessly about in the streets, helping the pariah dogs, and eating almost anything.

About Cairo, where it is considered a breach of order to kill them, they are called "Pharaoh's Hens." They are plentiful in Turkey, Arabia, and Persia, and are always to be seen about the camps and cantonments in India; but they also frequent the Alps and Pyrenees, and all the countries bordering on the Mediterranean.

Just as gulls follow ships, so do these vultures follow

the caravans across the desert, in the hope that something in the shape of a worn-out camel may turn up to their advantage.

It is said that during the war between France and England in the last century, the sharks learnt to know when a naval engagement was about to take place, and would assemble in the neighbourhood to be in readiness ; and in a similar way it was noticed that during the French occupation of Egypt, the vultures became so well acquainted with the meaning of the roar of artillery that they would flock together from all quarters as soon as the first gun was fired.*

Egyptian vultures are the most conspicuous birds in the island of St. Vincent, and outside the town of Porto Grande may be seen hunting over the heaps of refuse in company with ravens and crows, or lazily perched, half-a-dozen together, on the carcass of a horse or bullock which has been carelessly half buried in the shingle, or, more carelessly still, merely flung out of the town to pollute the air far and near until these scavengers have disposed of it. Having gorged themselves, according to custom, until they are

---

* "It is extraordinary," writes Frank Buckland, "how soon animals and birds find out where there is anything to eat. The regiment to which I belong very frequently marches down to Wormwood Scrubs for field-days. Upon arriving at the Scrubs I have not seen a single rook ; but the rooks very soon appear : they come to pick out what they can from the dung of the horses, and the bits of bread which drop out of the paper in which the men carry their refreshment. The rooks always go to the place where the regiment has dismounted, as there they find most to eat. These rooks come, I believe, from the trees in Holland Park—they certainly often arrive from that direction."

unable to fly, they will just flutter off a few yards when disturbed, but will not trouble themselves to do more.

The Aura vulture, popularly known as the Turkey-buzzard, from its resemblance to the farmyard-gobbler, is found wherever the country is moderately damp. On the coast of Patagonia it lives solely on what the waves throw up, dead seals and the like, and each herd of seals is sure to have a Turkey-buzzard watching it attentively. In hot countries it is a great blessing, and enjoys the protection of the American Spaniards, to whom it is very useful, since it haunts the slaughter-houses, walking about as tamely as the barndoor fowl, and eating up all the refuse. To the planter it is also welcome, and the day after the customary burning of the trash in the cane-fields, it is sure to be there feeding on the snakes, lizards, frogs, and other animals, which have been stifled by the smoke. It also did inestimable service, says Mr. Waterton, during the plague in Malaga, for it was impossible to bury the dead, and though they were thrown into the sea, many were washed on shore again, and when the wind blew landwards might have produced a second pestilence but for the vultures which came down from the hills.

The King of the Vultures, a bird with bare head and neck, which are coloured rich scarlet on both sides, is so called from the respect with which he is treated by the common vultures, none of whom seem inclined to begin their meal until he has finished his, though as many as a score may be present watching him, and will fall to eagerly when he has withdrawn.

In England the chief scavenger-birds, besides the gull,

are the raven, magpie, hooded or Royston crow, and the carrion-crow, which is, in fact, a small raven; but the last two are so destructive that in some places a price has been put upon their heads. The hooded crow frequents marshes near the sea, and the banks of tidal rivers, such as the Thames, where it may be seen within a few miles of London. In the western isles of Scotland flocks of five hundred may be seen in the month of June, and, like the carrion-crow, feed on dead fish and refuse of any kind, but also on living mollusks—as cockles, mussels, and the like,—which they drop from a height, in order to break their shells. ·

Crows are widely distributed in most parts of the world; the raven being the most conspicuous member of the tribe. It is widely known north of the equator, and is protected in Bengal, and unmolested in Egypt. It is an indiscriminate feeder, and while on the coast it subsists on dead fishes, in the polar regions it follows the herds of bison and reindeer, ready to take advantage of any that may be disabled by accident, or killed by wild beasts. No sooner has an animal been slaughtered by the huntsman than crows arrive in numbers. It is also a constant attendant at the fishing stations, and in North America, where it abounds, it robs the hunters' traps; and in the United States, whenever the deer are hunted without dogs it arrives to take part in the sport, and, not satisfied with what rightfully falls to its share, obliges the huntsmen to be very careful in concealing such game as they are unable to remove from the woods, as its scent is very keen.

The raven of Mexico and South Africa is of a different and larger species than that known in the north.   In South America it seems to be altogether wanting, but its place is well supplied by the numerous Caracaras and the Gallinazo.   The latter, called also Vulture Jota, Black vulture, Zopilote, Urubu, and carrion-crow, is of the size of a peahen, and, together with the Condor and King vulture, belongs to the small family of "flesh-bearded vultures."

The Gallinazo, though seldom seen on the Atlantic, north of Newbern, in North Carolina, is said to be found at Detroit, Lake Erie, and is very common in the south, where it ranges as far as Cape Horn.   Its preference is for a damp climate or the neighbourhood of water, and it abounds throughout the pampas, is preserved as a scavenger in Peru, and is generally protected by law throughout tropical America.

In many of the towns and villages of the Southern States it is as common as poultry, and may be seen sauntering in the streets, loitering indolently for hours at a time in one place, or sunning itself on the roofs and fences, and cowering over the chimneys if the weather be cold.   The townspeople, though disgusted by its filthy voracity, and particularly by a most unpleasant habit of disgorging its food down their chimneys sometimes, when it has eaten too much, yet respect it as a valuable scavenger, and, accordingly, it is protected either by law or custom.

Don Ulloa speaks of the Gallinazo as familiar in Carthagena, which it cleanses of all animal impurities, and calls it an "excellent provision of nature," for in that hot, damp climate the effluvium arising from putrefaction would be quite intolerable. In the town of Savannah it walks about in great numbers, devoting especial attention to the quarter inhabited by the hog-butchers. But it is said also to scent, or in some way to discover, carrion at a distance of three or four leagues. Its sight is certainly extraordinarily keen, for when compelled to search for food it rises to such a height as to dwindle down to a hardly visible black speck, yet from that vast distance it watches intently the movements of both animals and hunters, knowing well that the latter often kill a bison for the sake of its skin and marrow-bones, and leave the carcass for its benefit.

The Turkey-buzzard will wait and watch its food, not condescending to touch it until it is well seasoned ; but the Gallinazo is less fastidious. A horse having dropped down and died in the streets of Charleston, the carcass was dragged out to the suburb of Hampstead, where in a short time it was covered and surrounded by a dense crowd of these so-called " carrion-crows," many of which sat on the tops of sheds, fences, and houses, while several hovered in the air overhead, and at a distance. At one time 237 were counted, but probably this does not represent the whole number ; for the ground was simply black with them for the space of a hundred yards on all sides of the carcass, thirty-seven were upon and immediately around it, so that scarcely an inch of it was visible, and several were inside, presenting a most savage appearance as they from time to time emerged. Three or four dogs were assisting at the scene, growling and snapping when the wings of the

" crows " flapped them ; but though the birds sprang up for a moment they were not alarmed, and did not disturb themselves even when the dogs' master advanced near enough to order them home.

Though always bold, and generally protected, the Gallinazo does not seem to meet with quite such friendly treatment in Villa Nova as elsewhere, perhaps because licence has made it too troublesome.

It assembles in great numbers in the villages, Mr. Bates tells us, about the end of the wet season, and is then so ravenous with hunger that it is not safe to leave the open kitchen for a moment while the dinner is cooking. Some of the birds are always loitering about, watching their opportunity, and the instant the cook's back is turned, they will march in and lift the lids of the saucepans with their beaks. The boys of the village lie in wait and shoot them with bow and arrow, and the vultures have come to have such a dread of these weapons that a bow suspended from the rafters of the kitchen is often enough to keep them off. As the dry season advances, multitudes of them follow the fishermen to the lakes, where they stuff themselves with the offal of the fisheries, and when, towards February, they return to the villages, they are not nearly so ravenous as before their summer trips.

In South America, when the Gallinazo has begun the feast, the bones are picked clean by two species of Caracara, or Carrancha, which, from their structure, are called eagles, though in habit they resemble the carrion-crow.

The larger of these two, the Brazilian Caracara swarms in the desert between the Negro and Colorado, where it watches the line of road to devour the carcases of exhausted animals. It is most numerous on the grassy savannahs of La Plata, but also frequents the sterile plains of Patagonia, and although these false eagles rarely kill their prey, any one who falls asleep on the plains will see on awaking that he has been watched with evil eye by a Caracara perched on each hillock near him. Like the Gallinazo, it is to a certain extent domesticated, and constantly attends the slaughter-houses, while several will accompany a hunting-party.

The other species of Caracara, called Chimango, is much smaller, and is generally the last to leave a skeleton, and may often be seen, like a bird in a cage, running about within the ribs of a cow or horse. It is extraordinarily tame and fearless, haunts the neighbourhood of houses for offal, and if a hunting-party kills an animal, it soon assembles in numbers, standing on the ground on all sides and waiting patiently until its turn comes. It will readily attack wounded birds, and several together will even seize a cormorant.

It is a mischievous and very inquisitive bird, and Mr. Darwin mentions that it would not only tear the leather from the rigging and carry off meat or game hung up in the stern of the vessel, but would also pick up almost anything from the ground, and on one occasion carried a large black glazed hat nearly a mile, and on another took away a small compass in a red morocco case, which was never recovered. All the carrion-birds are commonly protected in tropical America, and so also in the eastern hemisphere is the Arabian kite, which haunts human habitations and pays a

Q

visit to every house in the village to which it attaches itself. At St. Jago, says Mr. Moseley, a flock of kites will come swooping about the ships to pick up garbage, which they seize in their claws with wonderful precision, putting out one foot to snatch the morsel and then bending their heads and eating it at once on the wing. This manœuvre is not always a safe one, however, as a shark will sometimes snap at the bird's foot and pull it under the water.

In India, the Bramah kites, or "Bromley kites," as the sailors call them, haunt the ships with similar intent, and are extremely bold. On one occasion, as a ship's steward was carrying a hot steak from the galley to the cabin, a kite swooped down on him, caught up the meat with its foot, and was off again in an instant, leaving the man bespattered with gravy. These kites also frequent the places where the Parsees expose their dead.

The Argala, or Adjutant, a bird of the stork tribe, has obtained the latter title from the fact of its being a constant visitor to the parade-grounds in India, and presinting a resemblance to the dress and dignified walk of the military officer of the same name. The bird, however, also condescends to make itself generally useful by cleaning the streets. This it does most thoroughly, for its appetite is large, and it can accommodate a full-grown cat or a leg of mutton without difficulty.

The great White Stork, about which the Germans and Dutch have so many pretty fancies, is a well-known summer visitor in many parts of Europe, and, besides consuming offal, helps to keep within bounds the swarm of frogs with which Holland would otherwise be over-run. It walks fearlessly about the streets and fish-markets, and builds its nests on the top of almost every pillar in the ruined cities of the East.

We must conclude this brief account of the principal scavenger birds by mentioning that the domestic Duck is as greedy a consumer of filth as any vulture, and does not shrink even from devouring her own deceased kindred when opportunity offers. Much the same may also be said of the Pig, which will eat garbage, and even carrion of any kind, and is therefore too often converted into a domestic scavenger by unthinking people, who put tainted meat, decaying vegetables, sour food, and refuse of all kinds into that often terrible receptacle, the pig-tub, and then expect their pig to convert its evil-smelling contents into wholesome, or at least eatable and saleable pork for them.

Rats are in general miscellaneous feeders, and, when pressed by hunger, will eat almost anything; but the Norway Rat, as it is called, which has nearly exterminated the old English black rat, frequents the premises of bone-boilers and knackers, as well as the sewers. Originally a native of India or Persia, it seems to have moved on into European Russia, whence it has been carried by merchant ships all over the world, and wherever it has been introduced has speedily ousted the native rats. The Black Rat lives chiefly in the ceilings and wainscots of houses, &c., does not affect such low places as pig-sties and cellars, and is but rarely found in the sewers, where the Norway rat swarms.

The rat, though one of the most despised and tormented animals, is yet, says Mr. Buckland, a most useful servant, for wherever man settles, there, as if by magic, the rat makes his appearance. Thousands of rats lived in the camp before Sebastopol, and they swarm at Aldershot, where the sentries see them at night going to the nearest water to drink, for the rat is a thirsty animal, and soon dies if kept without water.

The rat clears away every particle of refuse and filth he can get at, and does invaluable service not only in camp but in the sewers adjoining some of the London slaughter-houses, which are often nearly choked with offal and refuse animal-matter thrown into them by the careless butchers. But for the persecuted rats, who live there in swarms and devour every morsel, this putrid mass, if left neglected, would give rise to fearful plagues.

The rat is the only animal which can thrive and keep a clean coat in the most filthy localities where the air would be fatal to any other creature ; and, in spite of the unclean places he frequents, he is personally very particular about cleanliness, and never eats a morsel of food without cleaning himself; nor does the garbage upon which he feeds poison his teeth, as has been said, or render the wounds he inflicts with them deadly.

Rats have very sharp teeth, and are so fond of taking a nibble at the tip of an elephant's tusk that much of the ivory imported bears evidence of having been gnawed by them. Indian ivory they will not touch because it is deficient in animal glue, or gelatine, and of the African they taste only the best tusks, and of these only the purest and most delicate portion ; and the turner, well knowing that he may trust to their judgment, chooses a tusk which the rats have gnawed when he wants a specially good bit of ivory.

Mr. P. L. Simmonds mentions that rats are, or were, turned to account in another way in Paris, where there is a large pound, covering some ten acres of ground and surrounded by a stone wall, to which all dead carcases are brought. The bones of the animals are valuable, and so, of course, are their hides ; but they must be freed from the flesh, and how to get rid of this in a sufficiently expeditious, economical, and inoffensive way, was a difficulty, until some one suggested that rats might be employed. They were accordingly introduced by thousands, and did the work required of them to perfection, for a dead horse put in at night would be found turned into a neat and even polished skeleton by the morning.

Among field-scavengers must be reckoned the hedgehog, which feeds on animal and vegetable refuse, devours dead game which the sportsman has lost, and probably puts out of their misery such wounded creatures as have escaped the dogs and have crept into some hole to die.

But of all the mammalia, the most genuine carrion-feeders are the *canidæ*, or dog-tribe, and the hyænas. Though undoubtedly of great use, they are not perfect scavengers, inasmuch as they prefer their food tainted, and hence their own odour, like that of the vultures, is disgusting.

The pariah dogs are a feature of Eastern cities too well known to need much remark. In Ceylon they are not natives, but European mongrels, a most miserable race, having no owners, living on the refuse of streets and sewers, and subject, as the reward for their services throughout the year, to a general annual massacre to keep down their numbers.

As for the celebrated dogs of Constantinople (Fig. 50), to whom the scavenging of that city is mainly left, they are creatures with whom, as a traveller wrote to us a few

Fig. 50.—STREET DOGS OF CONSTANTINOPLE.

years ago, "all well-conducted sprigs of canine nobility would indignantly refuse to recognise any connection. They have four legs and a weakness for barking and biting, but here all resemblance to their European namesakes abruptly ceases. Here they are the lords of creation, and though in all conceivable stages of manginess and disease, are suffered to lie unmolested in the middle of the streets, and receive a sort of superstitious homage from their Turkish masters. Not indeed that they have any recognised owners; they are the fourth estate of the empire, and exercise, so at least it is said, a considerable influence on the sanitary arrangements of the capital. They have no fixed abode, at least in the eyes of the inhabitants, but nevertheless, among themselves, the city is portioned out into numerous and well-defined districts, and woe to the dog who trespasses on his neighbour's territory! His life, unless he beat a speedy and ignominious retreat, is not worth five minutes' purchase. Though held in veneration by the Turks, they receive no regular meals, and consequently are compelled to have recourse occasionally to an extremely sparse and delicate diet. I saw one to-day making his midday meal off a couple of cherries and a rain-puddle.

"In personal appearance they are between a fox and a wolf; in disposition, easily roused, snappish and cunning; in taste, omnivorous. Their dismal concert, which takes place every night, and is never postponed by reason of the inclemency of the weather or the sore throats of any members of the company, constitutes an effectual antidote to the sleep of those whose notion of music is not based on the growl and the howl."

The wild pariah dogs of India frequent the jungles and the lower ranges of the Himalayas in numerous packs, and whenever there is any fighting going on they are sure not to be far off.

All the scavenger dogs of Eastern cities, even the mongrels of Colombo, seem to a certain extent to follow the example of their Constantinople kindred, and divide the streets and lanes into districts. In 1844 the dogs of Alexandria had become so numerous and troublesome, that Mehemet Ali banished them in a body to an island at the mouth of the Nile ; but no sooner were they gone than their place was taken by a swarm of dogs from the suburbs, and the nuisance became so much worse than before, that the banished dogs were recalled and very speedily put the interlopers to flight.

Under the head of the *Canidæ* come both the Jackal and the Wolf, the former of which (Fig. 51), though he waits respectfully for the tiger to finish his meal, stands in no such awe of the vultures, with which he will dispute the remains of the carcases, snapping at them in defence of his rights. It is a very bold animal when pressed by hunger, and will

Fig. 51.—JACKALS.

not only follow the hunters and shamelessly take possession of their game, but will enter the streets of towns under cover of the night, eat the offal, and visit the hen-roosts, and even the larders, devouring everything it can find in the way of provisions, whether it be carrion or only cooked vegetables. It has also a taste for fruit, and a pack of two hundred will sometimes pay an evening visit to a vineyard.

The Prairie-wolf will also follow a hunting party, but much more cautiously than the jackal, and so suspicious is it that the sight of a stick planted in the ground, with a strip of calico fluttering from the top will be enough to prevent it from carrying off the dead game.

Of all scavengers, none is more horribly repulsive in appearance or more disgusting in its ways than the hyæna; yet Mr. Wood calls it the "very saviour of life and health" in Asia and Africa, and declares it to be a libel on the animal to say that it is incapable of being tamed. There are several species, and from the immense quantity of fossil remains, hyænas would seem to have been still more numerous in former ages, and to have been almost the sole scavengers of the great mastodon, mammoth, &c.

With its extraordinarily powerful teeth it does all the rough work of scavenging in the desert, in the forest, on the beach, crushing with ease such bones as would resist the strength of any other animal, and finishing up even the hides and other tough morsels left by them. It has been known to drag a dead camel above a mile from the caravan in the course of a single night, but probably in this case two or three, or perhaps more, acted together.

When they are too numerous to find sufficient carrion for their support, they become terrible pests, hanging on the outskirts of villages and encampments, even roaming through the streets at night and carrying off not only cattle but sleeping children.

It is a very cowardly creature, and is afraid to touch any animal unless it takes to flight, and thus the sickly often escape by standing still, while the strong fly and perish.

With its horrible voice, offensive odour, great personal uncleanness and cowardice, together with its habit of digging up the dead and attacking domestic animals and human beings where it can do so with safety, it is universally detested, notwithstanding its important services.

The *Felidæ*, or cats, great and small, wild and domestic, may be reduced by circumstances to eat almost anything, but since they prefer to catch and kill their prey, they cannot be considered as genuine scavengers, though the panther eats carrion when other food is scarce, and the lion of Algeria actually haunts the neighbourhood of towns, and satisfies his hunger with the garbage of all sorts flung outside the walls.

However, Nature's scavengers, though they unquestionably do a vast amount of good, by preserving the air from pollution and ridding the earth of that which is disgusting to sight and scent, cannot generally be called perfect. Some are disgusting in themselves, and others do so much mischief by their want of discrimination that many people are disposed to find fault with them, and to question their title to be looked upon as benefactors at all.

But, in the first place, surely Nature's workmen can be fairly judged only where they have to deal with Nature alone, and are not brought into contact with civilisation; and, secondly, if they transgress the bounds within which they are useful to man, whose fault is that?

CHAPTER XVI.

ANIMAL REMAINS AND ANCIENT DUST-HEAPS.

Animal remains chiefly Marine—Land Animals, how they may be Buried—Places to Die in—Birds seldom Buried—Wingless Birds—Guano-beds—Immense Supply—Coprolites—Rock-oil from Animal Remains—Shell-sand, Fossils, Casts, and Models—The Mammoth, "Giant-rat," "Grip-claws"—Bones Preserved in Stalagmite—"Kitchen-middens"—Cave-dwellers—Shell-heaps—Lake-dwellers.

SINCE the rocks forming the earth's crust have been deposited chiefly in salt water, as has been shown, it follows that the remains which they enclose will be mainly those of animals living in or near the sea. Remains of shells, corals, fish-bones, saurians, &c., are naturally abundant, and so are fish-scales, a modern deposit of which is to be found on the shore near Dundee, some ten yards long, and two or three feet thick. These are the scales of herrings, which fall off when the fishes are cleaned, and, being very buoyant, and comparatively indestructible, are thrown up by the waves.

The case of land animals is altogether different; for with the vast army of hungry scavengers always on the watch, no dead body is likely long to escape being devoured if it remain exposed, and the circumstances under which it is likely to be buried and preserved are exceptional. Old land surfaces have occasionally been buried beneath sediment, and where this has been the

case animal remains are abundant. At times they may
be covered by the mud and sand of inundations, at others
by the sand which drifts in from the sea-shore, and at
times animals are overwhelmed by landslips, or lost, and
that in considerable numbers, in bogs and swamps, while
in limestone districts they fall into fissures or wander into
caverns, where their bones may be covered with a crust
of stalagmite.

Mammal remains are most abundant in the sites of
lakes into which the animals were, no doubt, carried by
flooded rivers.

Sheep and cattle are often washed away even in
England in the spring-time, and where rivers are larger
snowfalls heavier, and changes of temperature more sudden,
large numbers perish at times, and may be carried away
to lakes, estuaries, or even the sea.

During the great drought in the Pampas several
hundred thousand animals rushed into the river Paraña,
and perished from lack of strength to crawl up the
muddy banks again. More than once the carcases of
above a thousand wild horses were seen together, and, as
floods followed, large numbers of skeletons were probably
buried in mud.

Many animals seem to choose spots to which they
retire to die. On the banks of the Santa Cruz there
are places which are white with the bones of the guanaco;
at St. Jago there is a retired corner, to which the goats
betake themselves; and every one remembers the elephants'
cemetery in Ceylon, to which Sindbad was conveyed.

A dead elephant is never seen in that island, nor are

its tusks or any portion of its skeleton found. The natives declare that the herd bury their deceased companions if these die before reaching the solitary valley to which they are supposed to withdraw on feeling the approach of death. Every one believes in the existence of this valley, though it is mysteriously concealed from human eyes, and Sindbad recognised it at once, when, on recovering his senses after his alarming journey, he "found. himself among the bones of elephants, and knew that this was their burial-place."

Birds, having no such burial-places, and being less liable to be buried alive than other animals, are less frequently found fossilised. But when flying near volcanoes in a state of eruption, they have often been observed to drop down, killed by the noxious vapours, and if buried in fine volcanic ashes not only their bones, but the form of their bodies, would be preserved. Moreover, they sometimes fall into lakes when chased by hawks, and those which build near water are sometimes surprised and swept away by a flood, and in these cases, if they do not float long enough to be devoured, which, being very light, they probably do, they may chance to sink, and be buried in mud.

The chief remains of ancient birds are those of the large wingless kinds, whose bones were filled with marrow, instead of air, which made their bodies considerably heavier, and more likely to sink, while their want of wings put them on a level with other animals.

The most extensive accumulations of organic matter due to birds are, however, the great guano beds of Peru, Bolivia, Africa, &c. Three-fifths of the guano is soluble, so that one year of English weather would be enough to wash away many of the deposits entirely ; and on the west coast of South America there is—owing to rain—no guano worth mentioning, except between latitude 13° N. and 21° S., while the locality in which it is most plentiful and most valuable is the rainless region of South Peru.

It is only recently that Europeans have learnt the value of guano, but under the Incas the birds were strictly preserved ; landing on the islands during the breeding season was forbidden on pain of death, and overseers were appointed to give their proper share of the valuable commodity to each claimant at the right time.

In the Cincha Islands, off Peru, the beds are two hundred feet thick, and the supply is almost inexhaustible. In addition to the droppings of countless sea-birds which have resorted to these spots for centuries past, the guano is also partly composed of the skeletons and eggs of birds, and the bodies and bones of fishes and seals. It has undergone much alteration by internal chemical changes, and emits a strong smell of ammonia ; and since it consists chiefly of phosphate of lime with soda, magnesia, and sulphur, it is a powerful fertiliser, and enables even the sandy desert around Lima to bear crops of maize.

Osite or Sombrero guano is brought from a small island in the West Indies, which is entirely composed of the bones of turtles and other marine animals, together with coral sand, &c., which have been cemented into a compact mass by the droppings of birds.

A magnificent crimson, called murexide, has been obtained from guano, while a fine purple has been found in the *copros* of serpents. The latter substance, which is very like plaster of Paris, was, and maybe still is, bought up at the rate of nine shillings a pound from the Zoological Society by a "doctor," or, perhaps, chemist. What he did with it was a mystery to most people, but no doubt he made the purpurate of ammonia from it.

The well-known fossils called "coprolites," consist chiefly of phosphate of lime, and received their name because they were supposed to be the fossilised droppings of huge saurians, or lizards, and other animals; but though some are no doubt true coprolites, and all evidently result from the decay of animal matter, they have generally lost all trace of organic origin, and are simply nodules of bone-earth, which when ground, or otherwise prepared, make a valuable manure. They are found in large quantities in the Suffolk Crag along with the bones and teeth of whales, &c., and are washed up in such abundance on the beach that people are constantly engaged in collecting them.

The beds vary from a few inches to several feet in thickness, and are found in Norway, the West Indies, Spain, and South Carolina, but were first dug in Cambridgeshire, where Dr. Henslow at once pronounced them to be "bone-earth, which," he said, "we are at our wits' end to get for our grain and pulse, and are importing as expensive bones from Buenos Ayres."

All animal matter contains a large proportion of carbon; and, as has been already mentioned, it seems probable that many deposits of rock-oil are derived from the remains of fishes, mollusks, crustacea, and the other minuter forms of animal life, of which many shales, limestones, &c., are largely composed. The silicious flags of Caithness, for instance, are impregnated with oily matter which is apparently due to the innumerable fishes embedded in them.

Shells and the like are some of the most indestructible of animal remains; but having already spoken of the vast accumulations of chalk and lime to which they have given rise, we need only add that the soft earthy carbonate of lime called "marl," which is formed of freshwater shells and occurs in layers and patches from one to several feet thick in bogs and old lake sites, was at one time dug or dredged up as manure for the pastures.

The importance of lime—and especially phosphate of lime—to many crops may be seen by a glance at the mineral composition of their ash.

Thus the ash of meadow hay contains 11·6 per cent. of lime and 6·2 of phosphorus; that of winter wheat 4·9 of lime and 7·4 of phosphorus; and that of red clover has 34·0 per cent. of lime and 9·9 of phosphorus.

" Shell sand " (consisting of shelly, coralline, and other limy *débris*) is often applied to clay soils, especially by the French, who value it highly; and though 100,000 tons are taken every year from Padstow Harbour, it is so abundant that much more might be used.

The name of "fossil" is given to all organic bodies, animal or vegetable, which have been naturally buried and more or less petrified or turned to stone.

Many fossil shells are, however, scarcely at all altered, and some even of the more ancient still retain not only their mother-of-pearl, but even their colouring. In others again,

as the animal matter or gelatine decayed, water containing
some dissolved mineral has filtered in and filled up all the
interstices, it may be with silica, or it may be with some
metal, and the shell is thus more or less mineralised.

Very often when the mud in which a shell has been
buried has become hard, the shell itself has been dissolved
away, and all that remains of it is a cast of the interior in
hardened mud or stone, and an impression of the exterior,
with an empty space between the two. This empty space
is again often filled with mineral matter, so that we have a
perfect cast of the whole shell inside and out.

The beds known as greensands are, it is said, largely
composed of the minute internal casts of foraminifera
(Fig. 28), whose tiny shells, before they dissolved away, were
filled with silicate of iron and potash.

About two-thirds of a bone consists of the earthy matter
(phosphate, carbonate, &c.) already mentioned; the remaining third is animal matter, a sort of gelatine, and as
this decays, mineral matter may filter in, and the bone
become petrified or mineralised. Both animal and vegetable substances may be mineralised to a certain extent in a
few weeks, or even days, when steeped in mineral water.

Occidental turquoise seems to be nothing more than
fossil bone or ivory, coloured by the infiltration of phosphate
of iron, while the true turquoise, which it resembles, but
does not equal, consists of phosphate of alumina coloured
by copper.

But there is another and more wonderful method of
petrifaction which is by no means uncommon, to which we
have alluded in Chapter XII. In this the shell, bone, or tree

R

trunk is neither mineralised by infiltration, nor merely represented by a cast, but the whole of its organisation is faithfully reproduced. A cast may give one a perfect idea of the appearance of any object, as seen from within or from without, but cannot show its structure ; whereas a model, which is what this sort of petrifaction produces, is an exact imitation. Each atom of the original substance is replaced by an atom of some other mineral. The most common replacement is that in which silica is substituted for lime, the former being, as we have seen, especially attracted by decaying matter. Strictly speaking, these fossils are not therefore organic remains, but perfect models of such remains.

Some few extinct animals are represented, however, by more than fossils, and more even than bones. Early in the present century, the first mammoth, still covered with flesh, hair, and wool, was discovered in the ice by a Siberian fisherman, who possessed himself of its two great ivory tusks, which he sold for fifty roubles, and left the carcass for the white bears and dogs to feast upon. Two years later, in 1805, the skeleton was still almost entire. The animal measured sixteen feet in length and nine in height, and from the long, stiff, black bristles and coarse red-brown hair and wool still remaining, it was evident that it was a species of elephant fitted to live in cold regions.

Long before it attracted the notice of naturalists, the mammoth had been known to the Siberian Ostyaks, who were so accustomed to finding the carcases buried and preserved in the frozen ground, that they firmly believed that the creatures lived there, and only died when they smelt the air. Its long curved tusks they considered to be movable horns with which it dug its way through clay and mud. The Chinese knew it in very early times as the "Tien-shu," or "giant rat," "a stupid, inert animal, which," they said, "avoids the light and lives in dark holes;" and some of their learned men thought these "earth rats" might be the cause of earthquakes, which they could not otherwise satisfactorily account for.

Even late in the seventeenth century, Father Avril, when travelling in Russia, was told that the ivory he saw was procured by men who ventured their lives in attacking the creature which produced it, which was as big and as dangerous as a crocodile.

The Arabs seem to have been the first to develop the trade in fossil ivory, and from the corruption of their word "behemoth" we get "mammoth."

Immense quantities of the bones and horns of the fossil rhinoceros are also thrown up on the shores of the Polar Sea, and the inhabitants of the Siberian tundras, or swamps, believe them to be parts of a colossal bird with which they declare that sundry persons have had terrific fights.*

The "claw of a griffon" was presented to Charlemagne, and the Russian merchants to this day never call the sword-shaped horns of the rhinoceros anything but "grip-claws." As gold sand is found in some places where these "claws" are buried, it is probable that the expression "taking gold from under the griffons," had its origin in this notion of the

---

* The vast region of the tundras extends from 64° N. lat. northwards to the coast. For nine months of the year it is covered with ice; in the summer it is a swamp, producing nothing but moss.

gigantic bird, which also may have been the ancestor of the gold-guarding dragons of fairy lore.

Nowhere are the remains of both mammoth and rhinoceros more plentiful than in the lowlands adjoining the Icy Sea. Multitudes are buried between the Lena and Kolimar, and one of the New Siberian islands is little more than a mass of mammoth bones, which have been worked for many years by the traders. One single sandbank has furnished the best harvest of tusks for eighty years past, and, in 1844, Siberian ivory, to the amount of 16,000 lbs., was sold in St. Petersburg. At least one hundred pairs of tusks are still sent to the market every year.

Many limestone caves contain large quantities of animal remains; but stalagmite is not such a good preserver as ice, and nothing is left but bones, most of which are broken, rubbed, rolled, or polished, as if they had been carried long distances by water.

Often, no doubt, the animals were surprised by sudden torrents, carried into the caves, and buried in mud, over which a crust of stalagmite afterwards was formed; but often also they lived and died on the spot where their remains are found.

At least 300 hyænas of different ages were buried in the Kirkdale cavern in Yorkshire, which contains also the bones of wolves, bears, birds, &c., and as all the latter are gnawed, while the former are not, we may conclude that the cave was once the home of many generations of hyænas, and that the other animals were only dragged there to be devoured.

In other caves, the bones of hundreds of cave-bears, wolves, lions, tigers, as well as of the mammoth, have been

Fig. 52.—Kitchen-Middeners and their Dwellings.

similarly preserved in stalagmite to tell us something about the animal population of Europe in past ages.

Other caves are, however, even more interesting than these, since they contain the " dust-heaps " or " kitchen-middens " (Fig. 52), not of animals, but of men. The cave at Carriga-gower, in the county of Cork, for instance, shows us that it was inhabited in ancient times by people who lived to a large extent on beef, mutton, and pork, but had no more idea than the hyænas of keeping their dust-heaps outside their houses.

They ate their food and threw the bones down on the mud floor, perhaps for their cats and dogs, whose remains are also found, and they stabled their horses in the cave with themselves. Sometimes they caught hares and rabbits, and they fed largely upon mollusks, especially limpets, periwinkles, and garden snails, whose shells are found in great numbers. The earlier inhabitants of the cave had stones for hammers and flint flakes for knives, but they were followed by others who were more civilised and possessed iron knives, an iron chisel, and a nail, and must have cultivated some sort of grain, since the upper stone of a quern or hand-mill has been discovered.

In Denmark there are old dust-heaps from three to ten feet high, and from 100 to 1,000 feet long, which contain implements of stone, horn, bone, and wood, frag-ments of rude pottery, charcoal, cinders, and the bones of many animals, some of which, such as the beaver, do not now live there.

On the coast of Peru, and a few miles inland, there are shell-heaps more than 180 feet high and above 300 in diameter, which have been preserved from decay partly by the growth of vegetation, and partly by a thick crust of carbonate of lime which has been formed on the surface by the action of the rain.

These great heaps contain charcoal, ashes, stones blackened by fire, the bones of fishes and birds, especially parrots, splintered human bones and stone axes, and are evidently the refuse remains of countless savage banquets.

Similar shell-heaps are found everywhere in the Fiji islands ; others, which are probably relics of the Bushmen, occur at the Cape of Good Hope, and contain limpets so large as to be good drinking-cups. In Australia the shell mounds left by the natives are so large and numerous that white men have worked all their lives at sifting out the undecomposed shells to be burnt for lime.

Crows, and even vultures, do something towards raising heaps of shells ; above a hundred of the former have been seen together feeding on mussels, and Mr. Barrow says that on one occasion, in a cavern at the point of Mussel Bay, he disturbed thousands of birds, and saw heaps of empty shells enough to fill some thousands of waggons.

The ancient races inhabiting the pile-dwellings, whose remains were first discovered about thirty years ago in some of the Swiss lakes, instead of piling their rubbish around them, which would have been inconvenient, threw it into the water, where enough has been preserved in the mud to give us a tolerably good idea of their manner of life. Piles were driven into the bed of the lake and connected by timbers, and upon this common platform each family had its own hut, with a trap-door in the floor for the convenience

Fig. 53.—Swiss Lake-Village of the Bronze Age.

of fishing, and no doubt also for getting rid of their rubbish (Fig. 53).

There must have been a large settlement on the Lake of Geneva, for the piles extend 1,200 feet along the shore, and 150 feet into the lake.   Many other colonies must have been of considerable size, for thousands of piles are found still firmly fixed in the mud, and at Wangen, on Lake Constance, more than 1,300 articles of stone, bone, and pottery have been recovered.

The huts were made of twigs, woven together and plastered inside with clay, and the inhabitants had not only abundance of fishes, but the flesh of stags, goats, wild boars, and foxes, which last they seem to have eaten in great quantities, to judge from the number of bones.

The earliest lake-dwellers had none but stone implements, and were contemporary with the elephant and rhinoceros.   Somewhat later we find that deer, wild boars, and wild oxen, were still abundant; and later still, a generation rose up which had learnt the use of metal, for their tools and implements were made of bronze, which is a mixture of copper and tin.   As tin was from the very earliest times brought chiefly from Cornwall, we must conclude that the people of the Bronze Period had some indirect dealings with the inhabitants of Great Britain, and, therefore, knew something of trade.   Their pottery is much finer in texture and more elegant in shape than that of their predecessors, and they had made other advances towards civilisation, having learnt to keep domestic animals, to eat beef, pork, and goat's flesh, to cultivate wheat and barley, and to weave cloth of flax and straw.   They even wore necklaces, bracelets, hair-pins, and "safety-pins" (Fig. 54), and they made a kind of bread of the "whole-meal" or rather "whole-grain" variety, the grains being roasted, slightly ground, and merely pressed into lumps.

From the seeds and stones yet remaining, we find that they had plums, raspberries, and apples ; and they also had hazel and beech nuts, whose shells, with all their other refuse, they threw into the water, never dreaming that centuries and even thousands of years after they themselves were gone, it would be brought to light and examined with the deepest interest by the learned men of a new era.

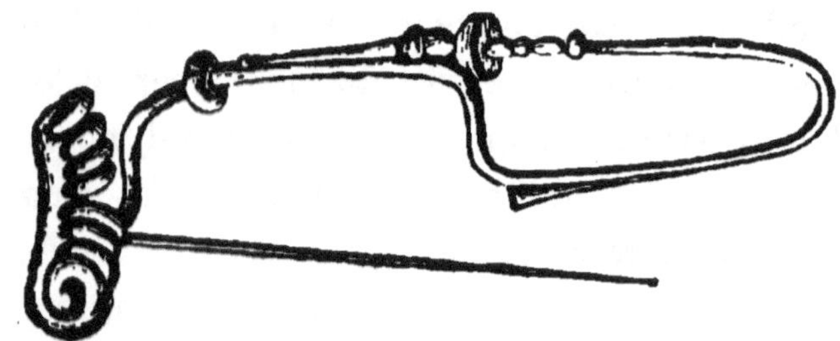

Fig. 54.—IRON "SAFETY PIN" FOUND IN LAKE NEUCHATEL.

· CHAPTER XVII.

HOUSEHOLD REFUSE.

Scavenging in the East; *Chiffonniers*; Different Classes; Their Dwellings; Hard Work and Honesty—Annual Sale at the Tuileries—Old Clothes, many Vicissitudes, what Becomes of Them—Old Uniforms; Old Hats; Rags—Woollen Rags—Various Materials used in Paper making—Waste - paper — *Papier-mâché* — " Hard and Soft Core "; Dust-sifting — Old Boots and Shoes, and "Jamaica Rum"—Broken Glass—Bones—Manure; " Bone-meal"; Bone-black—Old Pots and Pans—Broken Meat.

NATURE'S scavengers are not perfect, but the result of their labours on the whole is, that, where Nature is left to herself, not only is nothing wasted, but refuse of all kinds is so speedily disposed of that we seldom come across anything to disgust either sight or sense of smell. Where man comes on the scene, indeed, the case is different; but Nature can no more be expected to dispose of his refuse for him than to feed and clothe him, and provide for his other wants, though he oftentimes leaves a great deal to her, and owes her, as Miss Staveley reminds us, always far more than he is generally aware of.

Certainly, in the matter of scavenging, man does not compare favourably with Nature, for she knows nothing of the accumulations of filth and rubbish with which he is puzzled to deal; and, after watching her labourers at their work, it is rather humiliating to glance at a dust-yard and be compelled to own, as one must, that she manages matters in a much neater and altogether less offensive way than we do.

Throughout the East, scavenging, even in the towns and villages, is left chiefly to her good offices, as we have seen, and many a village on the hills about the Bosphorus owes such cleanliness as it possesses to the occasional heavy falls of rain.

In Ceylon, the Chandálas, who belonged to the lowest caste, were degraded to the office of scavengers and carriers of corpses ; and in the legends of Buddha, the chandála is represented as " one born in the open air, his parents not being possessed of a roof; he lies among the pots when his mother goes to cut firewood," and, says Manu, " he can never be relieved from bondage or emancipated by a master." In 437 B.C. the *nichi-chandálas* are specially named as " cemetery-men ;" but a couple of centuries later cremation had been introduced, and their services were no longer required in the cemeteries.

Not very much above the chandála in the social scale, come the Paris *chiffonniers*, who form a class as much apart from the rest of the community as if they were separated by the laws of caste. The *chiffonnier*, or " ragman," no more confines his attention exclusively to " rags " than the dustman does to genuine " dust." He is a by no means romantic or attractive-looking individual, and it is difficult at first sight to understand why he should be such a favourite character with French novelists, and how he acquired the halo of romance which surrounds him at the present day. According to the popular legend, every *chiffonnier* is a nobleman, or at least a gentleman, in disguise, or perhaps a learned professor who has met with reverses, or has other reasons for shunning the society of his equals.

But the *chiffonnier* would never have held this position in the public imagination had he, like the dustman, pursued his unsavoury occupation in broad daylight. It is his little dark lantern, and the fact that he does not emerge from his place of retirement until nightfall, which have combined to make him such an object of mystery and interest.

Like his Italian brother, and like most of the scavengers employed by Nature, he appears with the owls and bats, prowls about in the deserted, silent streets at the most unearthly hours, and when the sun rises he vanishes into his own quarter of the city, and is seen no more till night returns again.

We, who are accustomed to seeing the dust-carts on their daily rounds, may well wonder what the *chiffonnier* can find in the streets to fill his wicker-basket, and our wonder will increase when we learn that, according to the official returns, 7,050 men and women make their living in this way in Paris, and earn, some of them, as much as three francs a day.

The fact is, that as regards its scavenging, the elegant city of Paris has not hitherto been much in advance of Constantinople or Cairo. All the dust and refuse of the houses was simply thrown out into the street,* where the *chiffonnier* rummaged in it at his leisure with an iron hook, or a stick having a crooked nail at the end, carrying off

* This was also the case in Rome until of late years.

all that he could in any way turn to account, and leaving only the refuse of the refuse to be cleared away by the carts, or devoured by rats, hundreds of which may be seen running about the streets in all directions late at night. Of late, however, he has been thrown into great consternation by an order from the Prefect of the city, who considers that it would be more conducive to health and cleanliness if the "dust" of each house were collected in some receptacle, and put outside in the morning to be emptied into the dust-carts when they make their rounds. This plan is followed in Boulogne, and several British towns, and has great advantages over the system of dust-bins, since the refuse is removed daily, before it has time to decay and become injurious; and as the dustmen have merely to empty the tub or box, the "dust" of a whole street can be carried away in the time which it now takes them to empty a few dust-bins.

However, Paris has raised a great outcry against the new regulation; for, reasonable as it is, it means little less than ruin to the poor *chiffonnier*. When the "dust" is simply thrown into the street he can poke about in it easily, and soon pick out all that he cares to take; but if it were all collected in a deep tub, or such like receptacle, it would be long before he could ascertain what treasures might be hidden in its depths, and could make sure that he was not leaving anything valuable behind; and then, worse than all, the time for making these minute investigations would not be allowed him, for the "dust" would not be put out of the houses until the morning, and the carts would come and take it away before he had even a chance of completing his search. The *chiffonnier's* principal treasures are bones, rags, paper, scraps ·of copper, sardine cases and other tins, and broken glass; and of these, two hundredweights (about one hundred kilos) of bones bring him four francs; the same weight of paper from one to five francs, according to the quality; woollen rags, forty francs; copper, eighty francs; tins, three francs; white glass, six francs; and green bottle-glass, one franc twenty centimes, or about a shilling.* Cigar ends are also carefully collected, and converted into Régie cigarettes.

Money, jewellery, and the like, he is expected to give up to the *concierge,* or porter, of the house opposite which he has found it, and he bears a high character for honesty in this respect. The sardine and other tins, of which such thousands are thrown away, go to support a branch of industry which is deeply interesting to the youthful part of the population; for, after being stamped into shape by machines of simple construction, they reappear in the form of countless armies of tin soldiers, which are sold at prices so low that the manufacturer could not possibly afford to buy his raw material new from the tinman.

Rag-picking is a calling which any one is free to pursue, but still the fraternity have their rules and regulations, and are divided into well-marked ranks. Lowest of all are the *biffins,* or *chineurs,* who are half desperadoes and half amateurs, people who are something else by day, and go out fortune-seeking in a random, irregular way at night. They are not in favour with the other *chiffonniers,*

* A franc is worth 9½d. of English money.

for they either do not know or do not respect the customs of the society, and being apt to encroach on other people's rights, are looked upon as bunglers and dabblers, a "disturbing element," in fact, calculated to throw the well-regulated machinery of the body out of gear.

The *rouleurs* are more respectable than these, and have each their own particular districts, in which they make their rounds ; they do not trespass on their neighbours' property, and expect to have their own hunting-grounds treated with equal respect.

The *placiers* occupy the highest rank of all, and, as the aristocrats of the brotherhood, have the privilege of overhauling the "dust" before it is turned out into the street. The *concierge*, who knows them well, and treats them with benevolent condescension, allows them even to come into the house and pick the dust over undisturbed. A *placier* will visit from ten to thirty houses in this way; and in return for the indulgence shown him will fetch water, or do any other little service for the *concierge*. His place is worth something in a wealthy neighbourhood, and when he retires from the business he sells it to his successor, who makes his first appearance with him one Sunday morning in a tidy suit of clothes, and is formally introduced to all the *concierges* on his beat, and is recommended to their kind consideration. It is said that some *placiers* have paid as much as 1,000 francs for their business.

Besides the *chiffonniers* of these three classes, there are, above them again, the contractors, who buy the refuse of them, have it sorted by day-labourers, and then sell it again in large quantities. There are about a hundred such contractors in Paris, and of these four or five turn over millions of francs in the course of the year. Many are very wealthy, and all soon become well-to-do.

There is a certain M. Berton who has made himself conspicuous in the Paris elections since 1868, by offering himself as a candidate, printing his addresses on rose-coloured paper, and promising prosperity to mankind in general, and the Parisians in particular, if only they will give him a seat in the Chamber of Deputies or on the Municipal Council. This gentleman was for twenty years a dust-contractor, and retired from the business when still comparatively young with a yearly income of 60,000 francs (£2,040), earned for him by the *rouleurs* and *placiers*.

The *chiffonniers* live in certain streets, which are occupied by members of the fraternity only. The best known of these is the Rue Mouffetard, on the left bank of the Seine; but there are also large colonies at Montrouge, Montmartre, and La Villette, and everywhere their habitations look more like gipsy camps than anything else. Their huts are built of the most extraordinary materials, the walls being either of kneaded mud, or, like those of the ancient Lake-dwellers, of wattle, filled in with clay. A carriage-wheel often does duty for a window, the spaces between the spokes being covered with oiled paper, and the roof consists of tarpaulin, or, at best, of pieces of roofing-paper patched together. Some few buildings are constructed of brick, gypsum, or wooden spars; but these are owned only by the very prosperous. Great cleanliness is, of course, out of the question, considering the nature

S

of the goods which are piled up in and around the huts until they are sold; but the *chiffonnier* is personally less dirty than might be expected, and his health is as good as that of other poor people.

Most of the fraternity have been *chiffonniers* all their lives, for the calling is hereditary in certain families, and has been so for so many generations that the *chiffonniers* form a distinct caste, speaking, as Mr. Simmonds says, not a word of real French.

They have to work hard, making usually two journeys every night from the outskirts to the heart of the city. They have to turn night into day, and to carry heavy loads, and they must go out whatever the weather may be. But if their earnings are but moderate, they are, at all events, regular, and much more certain than those of many callings which rank far higher in the public estimation.

They are always on very good terms with the police, and though they may be said to be almost outside the pale of civilisation, are usually of no religion, and live and die without having any very certain civil standing, still crimes are almost unknown among them, and they are some of the most peaceable and orderly inhabitants of the unquiet city of Paris. They have never taken any part in the frequent revolutions, they kept tranquil during the Commune, and, it is said, went their nightly rounds and calmly picked their dust-heaps, even during that fearful week in 1871, when one part of the beautiful city was in flames, and Versaillists and Communards were fighting like savages in other quarters.

In England we have no one who at all answers to the

chiffonnier; but we have the " rag, bone, and bottle man," the " rabbit- and hare-skin man," the hawker who takes old clothes and " old 'ats " in exchange for pots of flowers and crockery ; and we are told that there are nearly a thousand persons who make their living by selling second-hand ar- ticles in the streets of London alone.

· Much might be said on the subject of old clothes, the largest dealers in which do so large a trade that they are known as " merchants," and export their goods to all parts of the world.

Long before the French Revolution, it was an established custom that there should be an annual sale at the Tuileries of all the discarded garments belonging to the Royal Family, the proceeds of which were ostensibly given to the poor. The practice was revived by the Empress Josephine, and continued through all the various changes of dynasty until the establishment of the present Republic.

A long gallery in the basement of the palace, looking into the garden, was fitted throughout its whole length with oak wardrobes which were usually well filled in the course of the year. When the time for the sale came, the shutters were closed, the gallery was brilliantly lighted up, and visitors were admitted by invitation-cards issued by the attendants of the Queen or Empress. Every article was ticketed with its price, from which, of course, there was no deviation ; but the chief part was generally bought up by the valets and women of the wardrobe, and they disposed of all that remained unsold to the great dealers, who again sold the goods to their customers at immense prices.

Many are the vicissitudes of old clothes, many are the

Fig. 55.—Petticoat Lane.

hands through which they pass, whether given away or sold, and with every change they drop a little lower in the social scale, becoming shabbier and shabbier, until at last, when they will hold together no longer, they are sold as rags, and forthwith enter upon a fresh career.

The trade in old clothes is almost entirely in the hands of Jews, whose great mart is in Houndsditch (Fig. 55). Some articles are sold just as they are, others are mended, patched, " translated," and made to look like new. The skirts of a coat, being the part least worn, are easily converted into children's clothes, but old black cloth always has a certain value even when too far gone to make miniature waistcoats or knickerbockers. France takes the best of it to make up into caps, and that which is still more threadbare is bought up for the same purpose by Russia and Poland. The black velvet waistcoats, still worn as best by certain classes, are converted into skull-caps for German and Polish Jews. The bulk of our old clothes goes to Holland and Ireland, but the vast majority of the scarlet coats worn by officers in the army are said to find their way to the annual fair held at Leipzig.

Regimentals, smart liveries, robes of office, &c., are greatly admired by the natives of the West Coast of Africa, and many of them are therefore despatched thither ; but the red tunics of the British infantry are chiefly bought by the Dutch, who make them into under-waistcoats which are worn next the skin by every careful working-man, since they are believed to be effectual in keeping off rheumatism, a matter of no small importance in their watery country.*

* Until recently, soldiers have been allowed to dispose of old uniforms as they pleased, and have been careful of their clothes accordingly, knowing

The grey overcoats of the infantry are sold at the Cape and in our own agricultural districts, where they are bought especially by the shepherds. The heavier and more valuable cloaks of the artillerymen are bought by the Dutch; and Holland and Ireland together are the purchasers of all the police uniforms.

A hat is an article which possesses a certain value as long as it will hold together and keep any shape at all; and it is wonderful what may be done with a battered old "chimney-pot," by means of cutting down, re-lining, brown paper, glue, rabbit-fur, dye, and varnish. By judicious treatment, the old hat may be made to look, for a time at least, like a smart new one, and as long as it is worn only in fine weather, the purchaser will no doubt congratulate himself on his bargain; but the first shower will reduce the whole fabrication to a shapeless mass, and he will probably regret the three francs which he has spent on the purchase. It is the French Jew who is especially clever at dealing with old hats, which he buys from the *chiffonnier* at eight sous a piece, and sells, at a considerable profit, for the sum above mentioned.

"Stockings! You can get as good stockings as *any*body wants for three-ha'pence a pair," says a poor woman living in Holborn. "They gives a penny a pound for stockings at the rag-shop, and sells them at a ha'penny a leg, and if you buys three legs you can make a first-rate pair o' stockings, good enough for any one!"

that the better their condition the better price they would fetch. At Aldershot the old-clothes dealers used to drive a very satisfactory trade, but disused uniforms have now to be returned, in accordance with an order to that effect, to the authorities.

But there comes a time in the existence of old clothes, whatever the rank of life in which they started, when they are too old, and patched, and faded, and threadbare, either for the dealers or the pawnbrokers. In fact they are no longer clothes but " rags," and in this condition they find their way to the rag-shops, either directly, or through the medium of the dust-cart.

It has been stated that not more than two-fifths of the rags in England are preserved, and manufacturers are consequently obliged to import large quantities, an expense which might be entirely spared to the country if people were more thrifty.

The London Ragged School at one time started some of its boys with four trucks, with the result that in nine months they had collected eighty-two tons of rags and other refuse, and 50,000 bottles. It is reckoned that each 600 houses would well supply one truck; but the Lancashire famine diverted all the rags in another direction, and the attempt to collect them does not seem to have been made again. In former days every thrifty housewife had a " rag-bag," into which were put not only rags, but all the snippings which result from the process of " cutting-out," and are now too often consigned to the fire. Some countries altogether forbid the export of rags, and our chief supplies are drawn from Italy and Germany, of which we are large customers.

All old woollen clothes come to the soil at last, being extremely valuable as manure. The early broccoli grown in the west of Cornwall thrive better on woollen rags than on anything else; and hops of a certain quality cannot, it is said, be grown without them. But good rags are so valuable for other purposes besides these, that it is chiefly the seams and other unusable parts which are put away to rot and then sent to the Kentish hop-fields.

But for the fact that old wool can be used again, all our woollen goods would be double the present price, for one-third of the manufactures of this country are supplied by " shoddy "and " mungo," that is, wool and worsted, no matter how old, which are reduced to their former state, and then re-woven.

In 1882 we imported 37,511 tons of woollen rags, at a cost of £820,616. Most of them are bought up by Dewsbury and two or three neighbouring towns, where they are torn to pieces by sharp spikes, and worked up into cheap cloth for the slaves of South America, or re-spun, with the addition of new wool, and manufactured into all kinds of woollen goods.

The " devil's " dust, as it is called, after the machine employed for this purpose, which is produced by the tearing up of the rags, befouls the whole town of Dewsbury, and is so injurious to the workpeople engaged in the factories, that they are obliged to keep their mouths muffled.

Some of the rags are ground to powder, variously coloured, and used for the making of flock-papers and artificial flowers; some are taken by the paper-mills and made into blotting and other papers of an absorbent character, and some again, after being boiled with pearl-ash, horns, hoofs, hoof- and hide-clippings, blood, old iron, waste leather, &c., reappear in the form of yellow crystals of prussiate of potash used in dyeing, and from this again the beautiful colour known as Prussian blue is manufactured.

Rags of all kinds are in great demand for the paper-mills, many of which use more than thirty tons a week. All are welcome, whether silk, wool, linen, cambric, lace, holland, fustian, corduroy, bagging, canvas, and even —though these cannot be called "rags"—old ropes. In fact, almost any species of tough fibre, even the roots and bine of hops, vine-tendrils, cabbage-stalks, and straw, may be made into paper. Straw alone makes the paper too brittle to be serviceable, unless the silica contained in it be destroyed, but it is frequently mixed with other materials. Still it is not probable that anything will ever lessen the value of rags, since they must needs go on accumulating and being disposed of in one way or another, and are, therefore, to be had cheaper than anything else.

Before they come to the mill it is necessary that the rags should be sorted, that the paper-maker may know exactly of what the bulk is composed and determine its destination accordingly, old rope being made into coarse brown paper, and the refuse of the flax-mills into tracing paper, while the paper for bank-notes is made from the best white linen. *

In 1882 we imported 21,200 tons of linen and cotton rags of very various degrees of cleanliness, those from Italy being lowest in the scale, while some of the English ones are said to be so clean as to require no bleaching.

* The tunics of the Jewish priests are said to have been unravelled and used as wicks for the lamps during the Feast of Tabernacles.

Rags have one advantage, besides their cheapness, over all other materials, in that the repeated washings which they have undergone while in use as wearing apparel, are an excellent preparation for their conversion into paper.

The first thing done with them is to open all the seams, take out any stray pins and needles, and remove buttons, which might injure the machinery or spoil the quality of the paper. Women are employed in this part of the business, and they stand before horizontal frames covered with very coarse wire-cloth, and having a large knife fixed upright in the centre, with the blade turned away from them. The bulk of the rags are cut up by machinery, but those intended for very fine paper are cut by hand into small pieces, about four inches square, by being drawn across the edge of the knife. Much of the dirt and sand passes through the wire-cloth into a drawer below, during this process; the remainder is beaten out by machinery, and the rags are then boiled with soda and lime. Clean white rags are said to yield from 65 to 80 per cent. of their weight in paper.

Statistics for 1884 show that there were in that year 3,985 mills in the world, which together produced nearly 17½ million cwts. of paper, of which the newspapers use about one-third.

But the rags have not come to an end of their career when worked up into paper, for the paper itself may be used over and over again, and clean waste-paper yields from 75 to 80 per cent. of its weight in new paper.

No paper, we are told, need be wasted, since it has been found possible to remove even the stain of printing ink, but still no doubt far more is consumed by cook and housemaid in the lighting of fires than is at all necessary. A whole newspaper seems to be considered a by no means extravagant allowance for one fire by some people, though it would light much better with one fourth the quantity, or even less. There are many shops where newspapers are bought at the rate of a penny a pound, which seems to be a protest against waste.

But it must be confessed that there is not much encouragement to people to save their waste-paper at present, in spite of the alluring advertisements offering to buy it up. In the country, where space is less valuable, room may be found for two or three sacks, perhaps, in an outhouse, and then it may be worth while to send them periodically to paper-mills, or other buyers. But in London one sack is about as much as most people can accommodate, and for this, two or three shillings, repayable when the sack is returned, have to be deposited; it takes a long time really to fill a sack, and when filled and fetched away it not unfrequently happens that a note is received saying that the value of the paper just balances the expense of fetching it, and this, even when the distance is less than a ten minutes' walk. People who have had such an experience as this, which is not uncommon, will certainly not care to take the trouble of making another collection, or to give up the space needed for the sack; but no doubt many would be glad to give away the accumulations from their waste-paper baskets, &c., if only some one would call for them periodically.

Meanwhile, however, large quantities of waste-paper do

find their way back to the paper-mills, to be re-made either into paper or *papier mâché*, in which latter manufacture no paper of any kind comes amiss. Tons of old account-books, bankers' books, and of ledgers; cuttings of bookbinders, of paste-board makers, of envelope- and pocket-book-makers, of print-sellers, of paper-hangers, &c., are disposed of for this purpose, and reappear as cornices, picture-frames, bulk-heads, cabin partitions, pianoforte-cases, chairs, tables, &c., not to mention the thousands of little fancy articles which are made of *papier mâché.*

There are three kinds of *papier mâché:* a sort of pasteboard, in which sheets of paper are merely pasted together, which is much used for tea-trays; a more solid kind, in which the paper is pressed until it becomes hard enough to take a polish; and a third variety, in which the paper is reduced to a pulp, in which condition it may be moulded into any shape that may be desired.

Paper, in the form of *papier mâché*, is daily being applied to fresh uses. Not long ago a daily paper announced that a factory chimney had been made of it. In America it has been used for the rails of railroads, the pulp being, it is said, as solid as metal, much more durable, and less influenced by atmospheric changes than iron or steel. Pasteboard wheels, made of equal parts of wood-pulp and straw, have been used exclusively for many years past by the Pullman Palace-Car Company.

Much waste-paper finds its way to the dust-contractor's yard, where it becomes the property of the "moulder," at least in some yards, for the regulations are not quite ·he same everywhere. The "moulder" is the foreman of the yard. He pays the sifters himself, and makes his own profits out of the paper, rags, bones, glass, iron, metal of all sorts, string, and corks.

The contractor has the hard and soft " core" and the coals and cinders. The " hard core "—consisting of broken crockery, earthenware, oyster-shells, &c.—is sold for road-making, but is now frequently unsaleable, and has to be got rid of at a loss. The " soft core," consisting of all sorts of organic matter, refuse from fish-shops, green-grocers, &c., is sold for manure, and the veritable "dust" or soil, which is as fine as gunpowder, is said to be especially useful for cultivating marsh lands and clover, and was at one time so much sought after by farmers that ship-loads of it were brought from the North. Some of it is also bought by the brickmakers to mix with clay.

Rags are not sorted in the yard, but sold in the lump to a Jew. The largest cinders and the coals are bought by laundresses and braziers, the smaller—called " breeze "—by the brickmakers, who use them as fuel, to burn between the layers of bricks.

In some yards the sifters are allowed to take wood, corks, and a daily quantity of cinders, as their perquisites, the allowance of the latter being so liberal that little markets are frequently held at the entrance to the yard, where the poor of the neighbourhood come to buy cheap fuel. In some places the women are also allowed pill-boxes and gallipots, and any crockery that can be matched and mended; and they may also appropriate the skins of dead cats. Cat-skins are worth from 4d. to 6d., the highest price being given for white skins. They are

used, not only as fur, dyed or undyed, but are made into
a sort of velvet, which has a very good appearance, and
cat- and rabbit-skins are also converted into felt hats. The
fashion of fur-tippets has in all likelihood considerably
raised the value of skins. Hare- and rabbit-skins do not
usually find their way into the dust-bin, but are bought
up by dealers in this particular kind of refuse. People
in England do not make nearly as much use of them as
their neighbours, and export two-thirds of them—that is
eight or nine million a year—to Germany, France, and
Belgium.

Hare-skins are much valued as chest-preservers, and
fetch from 18s. 6d. to 28s. the hundred.

Dust-sifting is necessarily a very rough and dirty occupation, but the wages are not bad, being from 8s. to 10s.
a week in money, besides the perquisites, and the health
of the dust-women is said to be exceptionally good. In one
large yard during the prevalence of cholera and smallpox
not a single person engaged there was even attacked. The
constant living in the open air, and the fact that they are
obliged to use a good deal of soap and water no doubt has
much to do with their good health.

Corks are re-cut or used in the making of kamptulicon
floor-cloth. In Paris old corks are collected from the
Seine, washed, re-cut, and sold at a few sous the hundred.
Waste cork, that which is too rough for cork-making, finds its
use as floats for fishermen and the stuffing of horse-collars.

Old boots and shoes go first to the " translators," who
patch them till they can be patched no more, and then
they are boiled down into glue.

A daily paper some time since gave curious statistics on the subject of old boots and shoes, which had been brought to light in the course of inquiries instituted in New York by the superintendent of the census. In New York and Brooklyn about three million pairs of old shoes are thrown away every year, and used to be plentiful in the gutters in some parts. They have now become scarce, however, as they are diligently picked up, and used for three purposes. Those which are not too far gone are patched and greased, and sold to the people who deal in such goods. Many folk wear one shoe much more than the other, and the dealers find pairs for the odd ones. The shoes not worth patching are cut up, and the good bits are used for patching other shoes, while all the worthless parts are converted into "Jamaica rum," by a process known only to the manufacturers. It is said they are boiled in pure spirits, and allowed to stand for a few weeks, and that the product far surpasses rum made in the ordinary way!

Broken glass of all kinds always finds a ready market, and may re-appear on our tables over and over again, for in making glass it is usual to melt the materials together with a quarter or half their weight of "cullet," or broken glass, of the same kind, so that many hundred tons are wanted in the course of the year. Of late, moreover, the coarsest kinds of broken glass have had another destiny opened up to them, being bought by a manufacturer—Mr. Rust—who melts them down, colours the paste any tint he chooses by a secret process of his own, and when it is cold breaks it into irregular fragments of various sizes, with which he produces very effective mosaics for the decoration of shops and other buildings, and, as the materials are cheap, he is able to sell them at prices much below those at which any other mosaics can be bought.

Broken bottles are also ground up to make glass-paper. The common kinds are made of the coarsest materials, of which rough sand and soapers' waste are some of the principal. Dirty water and soapsuds from laundries are used for watering gardens, or the grease is recovered and turned into soap again. The water in which fleeces have been washed yields fatty salts, called "suint," from which potash salts are made, and also soap, used in scouring woollen manufactures.

Bones are the only other considerable item which come under the head of "household refuse." In Russia they are either exported or simply "wasted," that is so far as man is concerned, being left to Nature to dispose of at her leisure. In more civilised countries they are put to a variety of uses, the majority of them going first to the bone-boilers to have the animal matter, oil and gelatine, extracted. Simply ground to dust, they form a valuable manure, and are imported in quantities from Australia, in the shape of bone-dust tiles. As "bone-meal," they are used for feeding cattle. At a large dyeing establishment in Manchester bones are boiled for the sake of the gelatine or size, which is used for stiffening goods, the fat is sold to the candle-makers, and size, liquor, and bones, are bought for manure. After boiling and bleaching—processes which render them more brittle than before—bones may go on to the turner, to be made into knife-handles, tooth- and nail-brushes, buttons, and the like.

Bone-black, or " animal charcoal "—as it is popularly but improperly called—is made by burning bones in closed vessels. From it is made the " ivory black " of the artist, and it is also used in the manufacture of blacking. Bone charcoal is employed for refining sugar, and is so absorbent that it will take all the colour out of treacle, or sugarwater coloured with indigo, leaving them quite white. The " charcoal " may be used over and over again by washing and heating, and when finally exhausted for refining purposes, is used for manure and the manufacture of phosphorus.

A word must be said about the old tin and iron ware—kettles, pots, and pans, &c.—which are sometimes consigned to the dust-bin. The tin-soldier business seems to be confined to the Continent ; in England the best parts of an ancient tin-kettle are clipped out, cut into shape, punched with holes, blacked and varnished, and used to strengthen the edges and corners of cheap trunks.

Old iron may, of course, be melted down, but before this happens it frequently takes a voyage as ballast, and large shiploads of frying-pans, gridirons, saucepans, candlesticks, tea-trays, boilers, shovels, old corrugated iron roofing, the produce of the old-iron shops, and the findings of the Thames mud-larks, are sent off to the United States and the Continent. Our exports of this description reached 132,033 tons in 1882, and were valued at £507,161.

Saucepans and frying-pans naturally lead one to think of food, and in concluding our notice of household refuse, we may mention a curious market held at the Halles Centrales, in Paris (Fig. 56), for the sale of broken meat of all kinds,

T

which was described at length by the correspondent of the
London *Daily Telegraph*, in 1878.

"The fragments which form the 'jewellery' of the
Halles Centrales, are brought down in big baskets between
seven and eight every morning by the *garçons* of the great
Boulevard restaurants, or by the *larbins* from the hotels of
the Ministers or the foreign Ambassadors."

If a grand dinner has been given the night before at
one of the Embassies, the show of "jewellery" in the morning
will be magnificent. There will be whole turkeys and
fowls, hams, and boars' heads, which have been scarcely
touched, displayed upon the deal boards.

Out of the season the supply comes chiefly from the
leading restaurants, where the "leavings" are the perquisites
of the waiters. Some dealers, and they are nearly always
women, have a yearly contract with particular restaurants;
some arrange the goods themselves; others, feeling that it is
a matter of importance, and that they do not possess the
requisite taste and skill, engage a professional hand to do it
for them. The object, of course, is to make a very little
seem a great deal, and also to render the various "portions"
as attractive to the eye as possible, and the "artists" flit
from stall to stall, giving here and there a touch of green in
the shape of spinach or Brussels sprout, or of red in the
form of carrot or tomato, adding a morsel of blanc-mange
here, a bit of pie-crust there, and so on, until each "portion"
looks as it should.

The "portions" are arranged on quarter-sheets of old
newspapers, and vary in price from two sous upwards.
As a sample, imagine a pile consisting of the leg of a

partridge, the remnants of an omelette, the tail of a fried sole, two ribs of a jugged hare, a spoonful of haricot beans, a scrap of filet, a cut pear, a handful of salad, a slice of tomato, and a dab of jelly, all to be had for five sous, or twopence-halfpenny !

The purchasers are, not the cheap eating-houses, as has been supposed, but the "quiet poor," people who are ashamed to beg, and but for the merciful cheapness of these appetising scraps would not taste meat from one month's end to another.

Fig. 56.—THE HALLES CENTRALES, PARIS.

CHAPTER XVIII.

MISCELLANEOUS REFUSE.

Utilisation of Refuse—Gas Companies ; Coal-tar and what is obtained from it—English Miners in Chile—Hydrochloric Acid—Soot and Smoke—Street-sweepings—Thrift of the Chinese ; Chinese Barbers—The Price of a Head of Hair—Refuse of the Fish Trade—The Queen's Tobacco-pipe—"Mahloo Mixture"—Cotton-seed Oil, Cotton-stalks, and Cotton-waste—Oil from Waste Products—Saw-dust—Metal Refuse—Furnace-slag.

WHAT one generation neglects, wastes, even pays to get rid of, another finds to be valuable property ; and a history of the utilisation of refuse would almost be a history of civilisation.

One of the most notable examples of the way in which the refuse of one generation makes the fortune of the next, is to be found in the history of the gas companies.

When gas has been separated from coal there remains in the retort, first, solid coke, which represents most of the carbon contained in the coal, and is so hard as to be able to cut glass, like its near relation, the diamond ; secondly, there is a certain amount of tarry matter and watery liquid. The latter, which is called gas-water, or ammonia-water, is a brownish liquid, with a strong smell, and yields salts of ammonia, from which spirit of hartshorn is made.

But the coal-tar ? Previous to 1856 it was worth hardly ½d. a gallon in London, while in the country the gas-makers were glad to give it away. Yet now, whereas nine million tons of coal at 12s. are worth but £5,400,000, the waste products on this amount, after the gas has been extracted, are, according to Dr. Siemens, actually worth £8,370,000.*

The fact is that the coal-tar which the gas-makers were so anxious to get rid of at any price, or no price, has been discovered to contain many most valuable substances, among which are the white crystalline acid called carbolic, now universally used as a powerful disinfectant; naphthaline, which crystallises in white pearly plates ; the colourless liquid called benzole, and white waxy-looking paraffine. All these are hydrocarbons, and as we have already seen that the essential oils to which plants owe all their sweet scents, are also compounds of carbon and hydrogen, it may not surprise us to hear that the perfume of woodruff, melilot, Tonquin bean, and many others, can be obtained from gas-tar, which also yields the peculiar odour of jargonelle pears —used in flavouring cheap confectionery—and oil of bitter almonds.

From benzole is obtained another colourless liquid called aniline, which is now manufactured on a very large scale, as, when mixed with other things, it produces the

---

* 9,000,000 tons of coal at 12s.	£5,400,000
Waste products on this amount :—	
Colouring matter for dyes	3,350,000
Sulphate of ammonia	1,947,000
Pitch	365,000
Creosote	208,000
Carbolic Acid	100,000
Gas-coke	2,400,000
	£8,370,000

brilliant colours extensively used for dyeing silk, woollens, and other goods, and for printing calicoes.

Not long ago the Corporation of Antwerp used to spend £1,000 a year in getting rid of the refuse of the town, whereas now they sell their street-sweepings and sewage for £40,000.

Some years ago, when English miners from Cornwall arrived in Chile, they were astonished to find the natives throwing away as useless the copper pyrites which they knew to be so valuable; this being the common form in which copper ore is found in Cornwall. But the Chileans were so certain that it contained no copper at all, that they not only laughed at the ignorance of the English, but sold them their richest veins for a few dollars. Then, again, the cinders from the old furnaces were thrown away as utterly useless, and it was actually found worth while to transport them to England, where, by stamping and washing, particles of copper were recovered in such abundance as to amply repay the purchasers.

Soda-ash, or sodium carbonate, is a substance manufactured in England on an enormous scale and used for glass-making, soap-making, bleaching, and various other purposes. Formerly it was prepared from barilla—*i.e.*, the ashes of sea-plants—but now it is obtained from sea-salt or sodium chloride, which is a compound of the metal sodium with chlorine gas.

In making soda-ash, sulphuric acid is poured upon the salt, and the sulphur combining with the sodium forms sodium sulphate, or "salt cake," which is the first step in the manufacture. But what becomes of the chlorine? It unites with the hydrogen of the acid,* forming hydrochloric acid gas, or spirits of salt, which passes away in the shape of a white smoke. That is to say, it would so pass away if it were allowed to escape, and it did pass away at one time, to the annoyance and injury of all who lived near the factories, for the fumes are very powerful, and not only destroyed trees and grass and every green thing in the neighbourhood, but ruined the farmers and gardeners for some distance. Then taller chimneys were built to carry the fumes higher up, and so they did ; but the gas was not destroyed, and simply came down again a little farther off than before, but with equally disastrous effects. This kind of thing, of course, could not be allowed to go on, and the next plan tried was turning the gas into the canals, where it was speedily absorbed by the water. But then there soon arose another outcry, for the people who owned barges complained that the gas made the iron rivets and nails come out of their boats. Finally, instead of being allowed to pollute the air or the canals, the gas was collected in a small quantity of water, and the Alkali Works Act was passed, compelling the manufacturers to condense it all and not suffer any to escape. And then it was discovered that this noxious matter might be converted into a valuable servant.

Hitherto the only way of bleaching known had been by means of the sun and air, but now chlorine gas was found to be a very powerful bleacher, and was made to do the work instead. "By passing hydrochloric gas and air together over heated copper sulphate," the hydrogen of the

* All true acids contain hydrogen.

one unites with the oxygen of the other to form water, and the chlorine is set free. Being wanted, however, in the form of a solid, not a gas, it is sent into lime, with which it forms the well-known bleaching powder, chloride of lime, also much used as a disinfectant, which is worth £17 a ton.

Two hundred thousand tons of common salt are annually consumed in Great Britain for the preparation of nearly the same weight of soda-ash, of which the value is about £2,000,000 ; and more than 1,000 tons of impure hydro-chloric acid are produced every week in South Lancashire alone during the process, all of which was, until of late years, not merely wasted but allowed to be positively destructive. The sulphur used in the process is still a waste product, and from its bad smell is a great nuisance ; but no doubt in time some use will be found for this also.

Another waste product which at present is allowed to be a general nuisance is the unconsumed carbon, which escapes from our chimneys in the shape of finely-divided soot, or smoke. The weight of soot in the air on a winter day in London is estimated by Dr. Siemens at fifty tons, while of poisonous carbonic oxide there are five times as much, and the two together " destroy public monuments, waste life, sight, and cheerfulness," and deprive us of so much warmth. Mills, furnaces, factories, bakehouses, &c., are now obliged to consume their own smoke, and all that is needed in private houses is some means of subjecting the smoke to heat sufficient to consume it before allowing it to enter the chimney.

It is said that smoke is a great disinfecting agent in populous towns, and as such it may be considered useful ;

but the same end might surely be attained in a civilised
community by other means without involving the present
drawbacks of wasted time, smarting eyes, injured bronchial
tubes, and defaced buildings. And what a wonderful
difference it would make to the spirits of Londoners in
general, to say nothing of the unfortunate artists, if the
air were clear, the sun allowed to shine upon them whenever he would, and they themselves freed from the necessity
of waging an incessant and more or less hopeless warfare
with "blacks!" Nature consumes her own smoke, why
should not we? Certainly the man who contrives to banish
it from our towns and cities will deserve all the fortune he
may be able to make out of it.

Soot, we may remark, is looked down upon by "dust;"
nevertheless, in addition to what we pay to be delivered
from it, soot is worth about 6d. a bushel, or a bushel of
soot is reckoned equal in value to a quartern loaf. Its
chief value as a manure arises from the sulphate of
ammonia it contains, but the notion that the soot obtained
from kitchen chimneys is superior to any other, owing to
the fatty matters mixed with it, does not seem to have anything to warrant it. Grass which is manured with soot
assumes a brighter green, and is much relished by cattle.
At one time it was exported to the West India sugar plantations; and besides being applied to the soil, it is used for
the manufacture of the brown colour called bistre, and for
the colouring matter of paper-hangings. Wood-soot is said
to be useful in hysteria and whooping-cough.

If we cannot at present collect and make use of the
refuse which pollutes the air, we are at all events less

helpless in the matter of that which accumulates in the streets, though even here our arrangements are far from perfect.

Street-dirt, or "slop" as it is technically called, when drained of its moisture, is sent off in barges to the brick-makers, who live a few miles out of London. They pay nothing for it, and even receive it carriage free, because the contractor is obliged to get rid of it in one way or another, and there is no one on the spot to take it off his hands, though in the country the farmers would be glad to pay for it and fetch it at their own expense, to put on the fields. Most of the street manure is now separated from the other sweepings, being collected by boys who dodge in and out among the horses and vehicles with great agility.

But of all people on the face of the earth, the most thrifty in the matter of dealing with refuse certainly seem to be the Chinese, who waste not a scrap of any sort, and cultivate every inch of ground, as, indeed, they have every need to do considering the millions for whom they have to provide food. Thrifty they are, but *nice* they are not, either in the original or acquired meaning of the word, at least in European eyes; and the odours which pervade their towns are terrible. They have no drains, and sewage and filth of all descriptions, together with every bit of organic matter, which cannot be used for food, even by a Chinaman, are put into the large tubs which stand along each side of the streets, at intervals of a few feet. The contents are used for manuring the fields, and no Chinaman, it is said, thinks of returning home from an expedition to the town without filling the buckets which he carries slung at each end of a bamboo.

Chinese barbers sell the hair of which they relieve their customers for manure; and a celebrated London barber told Mr. Buckland that, though now obliged to burn the cuttings of hair to get rid of them, when he was an apprentice in a country town the sweepings of the shop were allowed to him as his perquisite, and he was in the habit of selling them at sixpence a bushel to a farmer, who said that the land thus manured with hair required nothing more for three years. In London, where people keep their hair short, it would take a long time to collect a bushel of clippings, but country customers often have long locks to part with.

Human hair, by-the-bye, is the strongest fibre known, and a rope made of it was shown in the Japanese court of the International Exhibition of 1862.

The finest tresses used by hairdressers for making up into plaits, wigs, &c., come from the sisterhoods, and Paris seems to be the headquarters of the trade, as much as 140,000lbs. being sold there in the year. Many years ago pedlars, who went about the country in larger numbers than they do now, were in the habit of buying up hair, and have induced many a village lass to part with her beautiful locks for a trifle. One girl we know of allowed her luxuriant golden hair to be shorn off close to her head, and was satisfied to receive in return a brass thimble and a reel of cotton! In these days, or rather a few years ago such hair would have been worth a good deal.

But to return to the subject of manure. Dried hop-bines are found useful for this purpose, as indeed one would expect, and the "shells" of crabs and lobsters, which accumulate in regular mountains where the canning business is carried on, are ground to powder and applied to the soil with some success.

Better than lobster-shells, however, is the refuse of the fish-trade, large quantities of which are at present wasted, and might be had for the mere cost of collecting, as it is of little or no value. In cleaning cod for salting and drying, at least one half of the weight of the fish is thrown away, to be food either for the gulls and other birds attendant on the fishermen, or for other fishes if thrown back into the sea; and as the French and Americans alone catch some three million hundredweights of cod between them, the refuse must be enormous.

At least 50,000 tons of animal matter must remain, too, after the seals have been deprived of their oil and skins. On some of the North American coasts the offal is simply burnt or thrown into the water as food for the bait-fish.

Some years ago even the livers of the cod were of next to no value in Newfoundland, the people not having the necessary appliances for extracting the oil. But the arrival of Mr. Fox, an English chemist, soon caused them to rise in price, and he also made known the value of the heads which till then had been thrown into the sea or upon the manure-heap. Mr. Fox obtained from them a large quantity of superior isinglass as well as glue.

At the Exhibition of 1862 M. Rohart showed samples of fish manure from the Loffoden Isles, where he had collected heads and backbones, formerly wasted, and after drying them on the rocks in the wind, and subjecting them to other processes, had reduced them to powder. He had also bought up the half-used livers, extracted the remainder of the oil, and converted the residue into manure.

Boatloads of fish are often sold to put on the land at places on the coast, when the "take" happens to be unusually large, and there is no other more remunerative market for it.

Among the various kinds of refuse used as manure must be mentioned the damaged goods confiscated at many of the docks, which are buried until partly rotten before they are sold for this purpose. At the London Docks, however, the goods are burnt and reduced to ashes, many tons of which are sold to the farmer. In the centre of these docks a fire is kept burning night and day, its chimney being known as the "Queen's tobacco-pipe," and here are consumed all condemned goods, some of them damaged and unfit for food, but many of such value that it is deeply to be regretted that some more rational method of disposing of them has not been devised.

Great loads of tobacco and cigars are burnt from time to time, and a similar fate on one occasion befel thirteen thousand pairs of French gloves, while on another nine hundred Australian mutton-hams were condemned to the flames. These hams had been warehoused on their arrival, in the expectation that the duty on them would shortly be taken off, but they had to wait so long that they became damaged, and were condemned as unfit for food. No doubt their ashes helped to enrich the fields, but the hams might have fed a large number of people, so that we can look at their consumption in this way only as a very wasteful proceeding. Even in their damaged state some were perfectly eatable, and many a slice was eaten by the man in charge of the furnace.

Tea is now seldom burnt, having once set the chimney on fire; but cargoes of tea are sometimes condemned. Any one walking down the Mahloo road in Shanghai will see, on either side, trays of old tea-leaves drying in the sun, and exposed, not only to the dust, but to the attentions of the pigs, dogs, and children, which play and walk about among them. Many tons of this "Mahloo mixture," as it is called, are made and exported, and some of it is occasionally seized in London, and properly condemned, for no one but the consignees could possibly grudge it to the fields.

No doubt the waste from tomato-canning went for manure until an enterprising American conceived the idea of turning it into tomato catsup, from which he realises handsome profits, as he pays nothing for his material. The wholesale houses are, no doubt, glad to be rid of it, and to them he sends clean tubs to receive all the skins and parings. The tubs are removed every day, and the contents ground up, fermented, and flavoured.

Jute refuse is another of the substances formerly considered fit only to rot and be applied to the soil, whereas now it is mixed with flax, hemp, and silk, and made into excellent paper. And a somewhat similar history may be told of the cotton-seed, which was left to form offensive accumulations in some places, and in others given in small quantities to the cattle, or simply thrown away, or used as manure. "Throwing away" and "using as manure" are terms, by-the-bye, which very often, though not always, mean the same thing in the end; but the cotton-seed had a future before it, and what was a nuisance in 1824 was valuable property some years later; for in 1881 it was imported by England to the amount of 232,199 tons, the value of which was £1,783,109. It had been discovered, in the meantime, that cotton-seed could be made to yield nineteen per cent. of almost black oil, and that the residue made good "cake" for cattle. When refined, the oil is not inferior to fine colza, and is worth £29 a ton, but increases yet farther in value after it has made another journey, this time to the olive-growing districts of the South of Europe, where it is "doctored," and then re-shipped to England as "pure olive oil." Being almost tasteless, as well as nearly colourless, it is largely used for frying fish, packing sardines, &c. The inferior part of the oil yields hard grease, or stearine, which is employed for soap-making, and the husk of the seed is useful in paper-making, as well as for "cake."

Cotton-stalks are now turned to much better account than formerly, a manufacturer of Brooklyn, New York, having perfected and patented a machine for reducing to fibre any material of a fibrous nature. This machine is now extensively worked in New York, and the cotton-stalk fibre, originally intended for bagging, is now found to be much too valuable, and, being almost like hemp, is used for better materials.

The waste thrown out by the machine—the sticks and chips from the stalk—is more valuable as pulp for the manufacture of paper than any other substance yet discovered, and when mixed with cotton-seed makes nourishing food for cattle, or good manure.

In America, no doubt, the supply of cotton-stalks is abundant, but in Europe it would be a great boon to the manufacturers could they find some fibre which might be used with or instead of flax, and attempts are being made in Germany to utilise the nettle for this purpose.

" Nettle cloth" is still the German name for muslin, and nettle fibre was largely used in the olden days, before the introduction of cotton, as witness the nettle linen sheets and tablecloths used in Scotland early in the present century, and mentioned by the poet Campbell. What has been done once may be done again, and if manufacturers could be found to buy the fibre or stalks, no doubt many a waste piece of ground might be made to grow nettles with advantage, for it is a crop which requires little care, and is said never to fail.

Cotton-waste, which used to be employed for paper-making, is now made into wadding, lamp wicks, common carpets, and twine, it is also used for cleaning machinery, and is no longer burnt or thrown away, even after it has served this purpose, as was formerly the case, for it has been found possible to clean it, and both the cotton and the dirty oil, with which it is saturated, may be used again.

The lubricating oil used for machinery is also now regularly collected and cleaned. While on the subject of oil, we may mention that petroleum casks are collected by costermongers at 4d. a-piece from the oil-shops, taken to the water-side, and sold for re-shipment to America.

Among the other waste products from which oil is obtained may be mentioned grape seeds, which, though for the most part still wasted, have been long used for this purpose in Italy, the Levant, and part of Germany. Two casks of seeds yield thirty-three pounds of oil, which when purified is equal to olive, while the residue makes good soap, and the refuse of the seeds is used as fuel. The seeds are also valuable for fining and strengthening wine.

Argol, a crude variety of cream of tartar, which forms a crust in wine-vats and bottles, is also obtained from grape skins and refuse grapes, and from it is made tartaric acid, which is largely used in calico-printing, as well as for making lemonade. When the argol has been extracted the remaining refuse is made to yield gas and coke.

But there are still other sources of oil which must not be passed over. Orange peel—of which enormous quantities are wasted, *i.e.*, cleared away with the street-sweepings—contains much oil, which is easily expressed, and often used for making soap and scent. At some of the theatres and music-halls, where oranges are largely consumed by the audience, the peel is collected and sold.

Sawdust also yields oil, as well as spirit, oxalic acid, charcoal, and potash ; and from the oil and potash together, soap is obtained. Immense quantities of sawdust are produced in the great saw-mills of the United States, Canada, and Norway ; and except being used for packing, for sprinkling floors, and for smoking fish, as litter for horses, and for mixing with fish refuse as manure, nothing more used to be done with it. The sawdust is, however, much more valuable, now that the various substances above mentioned can be produced

U

from it, and at one mill in Norway two horses are constantly employed in removing it. From 9 cwts. of sawdust, over $19\frac{1}{2}$ per cent. of grape sugar is obtained, and from this brandy is manufactured. But the Paris cabinet-makers have invented a way of using the sawdust itself. By subjecting it to enormous pressure and intense heat they convert it into a solid mass, which can take a brilliant surface, and which they call *bois durci*, or tough wood.

The dust of mahogany, birch, and rosewood, is used for cleaning and dressing furs; boxwood-dust for cleaning jewellery; and the shavings made in the shaping of cedar pencils yield otto of cedarwood, in the proportion of 28 ounces to the hundredweight.

The dust made by ivory-turners is sold at sixpence a pound, and when boiled down makes the finest and purest animal jelly. Ivory-dust jelly was at one time a fashionable remedy in cases of weakness, and Mr. Buckland was of opinion that it only needed to be tried to convince people of its virtues. There is at all events nothing objectionable about it, which, is more than can be said for the " red-currant " jelly which as well as rum, the Americans are declared to fabricate from old boots !

Dye-wood residues—that is, wood shavings from which the best of the dye has been extracted—give a solution which is useful for tanning and also to a certain extent for dyeing; but the insoluble remainder was worth but a few shillings a ton, and was in most cases thrown away—the only use to which it could be put being to burn as fuel when mixed with tar-refuse—until a French chemist began to convert it into paper pulp.

Rough kinds of grey and brown paper are also made from the shoots of hops left after the cutting down of the bines ; dried hop-bines are a valuable manure, and even the spent hops, those which have been used in the making of beer, are found, when dried, to make good litter, and are said to improve the health of the horses.

The rest of the brewer's refuse, namely the spent malt, called "draff" or "dreg," being more than the cows could eat, was in Edinburgh thrown into the Leith, until an outcry was raised against the practice, when the brewers found that by pressing the "draff" into cake they could sell it to farmers at a distance, and put £60 a week in their own pockets. During the cattle plague, however, it was again thrown on their hands, and for a while they paid to get rid of it, until it was found that by drying, it could be converted into good food for horses.

Glycerine is another of the now valuable waste products once looked upon as worse than useless, and thrown away. It is contained in most oils and fats, both animal and vegetable, and is formed during the process of soap-making. It is now largely used in medicine, for the production of syrups, &c., for extracting perfumes, making confectionery, and as a preserver. In the United States it is also used for charging gas-meters, two million pounds being thus consumed annually.

In conclusion, a few words must be devoted to metal-refuse, and as the fashion of wearing crinoline periodically threatens to become general, we may begin by mentioning that crinoline-steels are an awkward kind of refuse, with which no one cares to deal. The *chiffonniers* reject them, the dustmen do not like them, and when the fashion was declining some years ago it is said that numbers were thrown into the streets of New York and other towns, where they were not only a nuisance, but dangerous, both to foot-passengers and horses. Yet enormous quantities of steel must be used in their manufacture, and this, of course, could be melted down.

The monument to Horace Greeley, the founder of the *New York Tribune,* is cast out of many thousands of pounds of old type, contributed by the printers of the United States.

The scrap iron, or rather steel, left over from needle making, being of the finest quality, is used for making gun-barrels; the waste from the steel pens made in Birmingham is sold to Sheffield, at £10 a ton (the original price having been £50 or £60), and is there re-melted. Steel filings are bought by chemists for the manufacture of steel-wine.

Old ship's copper and copper-scraps of all kinds are first converted into oxide by heat, and then dissolved in sulphuric acid, forming sulphate of copper, which crystallises in large blue crystals, and is commonly known as blue vitriol, which is largely used in calico-printing and the manufacture of various pigments, such as Scheele's green.

Copper pyrites, or sulphide of copper, is used for the same purpose, and as small quantities of gold and silver are frequently associated with this ore, the residue from the pyrites kilns in vitriol factories is found to yield both these metals as well as copper. Gold and silver to the value of £3,232 have been obtained from 16,300 tons.

Silver nitrate, or lunar caustic, as it is called when cast into sticks, is, as is well known, used in photography; and from defective pictures, clippings, sweepings, and washings, one firm has recovered two bars of silver, worth £44.

A carpet which covered the floor of one of the rooms in the mint of San Francisco for five years was, when taken up, cut in small pieces, and burnt in pans, with the result that its ashes yielded gold and silver to the value of 2,500 dollars.

From the silver found on the copper-sheathing of a ship it has been calculated that there must be about 200,000,000 tons dissolved in the waters of the ocean; while of gold it is thought there must be one grain in every ton of water. Both these metals must, we suppose, have been carried into the sea by the rivers, but no practical way of recovering them has been invented.

We have already noticed the large number of particles of steel and iron in the dust of railway carriages; but though less obvious, there must also be great quantities in the dust of our streets, worn from wheel-tires, horse-shoes, and the nails in our boots. Some day, perhaps, it will be found possible and profitable to extract this metal, before the mud is sent off to farmers and brick-makers.

In smelting impure ore some other mineral is melted with it, to form what is called a flux. Thus iron ore is melted with limestone, and the lime combining with the silica of the ore, sets the iron free. The silicate of lime thus produced is called slag, and great hills of it are often to be seen in the neighbourhood of smelting furnaces. Sometimes it is crushed and used for concrete walls, but large quantities—now lying idle—might be employed for making bye-roads, for foundations, macadamising, and as railway ballast.

Of late, however, it has been applied to a novel purpose. Air is blown into it while it is in a molten state, and the effect is to draw it out into exceedingly fine fibres, as fine as the finest·spun glass, and as soft as cotton. It is, in fact, a species of glass, though no one would guess it from its original appearance, and it is used for covering boilers.

In the Sandwich Islands the birds make their nests of a similar substance, made not by any artificial means, but by the great volcano, Mauna, which churns its molten lava so violently that the spray is dashed high in the air, and, cooling as it falls, forms threads of fine-spun green glass, called by the natives " Pelé's hair," which drifts away with the wind, and hangs in masses about the trees and rocks.

PRINTED BY CASSELL & COMPANY, LIMITED, LA BELLE SAUVAGE, LONDON, E.C.